CAMBRIDGE LIBRARY COLLECTION

Books of enduring scholarly value

Earth Sciences

In the nineteenth century, geology emerged as a distinct academic discipline. It pointed the way towards the theory of evolution, as scientists including Gideon Mantell, Adam Sedgwick, Charles Lyell and Roderick Murchison began to use the evidence of minerals, rock formations and fossils to demonstrate that the earth was older by millions of years than the conventional, Bible-based wisdom had supposed. They argued convincingly that the climate, flora and fauna of the distant past could be deduced from geological evidence. Volcanic activity, the formation of mountains, and the action of glaciers and rivers, tides and ocean currents also became better understood. This series includes landmark publications by pioneers of the modern earth sciences, who advanced the scientific understanding of our planet and the processes by which it is constantly re-shaped.

Memoirs of William Smith, LL.D.

William Smith (1769–1839) was best known as the author of the *Map of the Strata of England and Wales* – indeed, he was known as 'Strata' Smith. His *Memoirs*, edited by his nephew John Phillips and published in 1844, tell the story of his life from his beginnings as a blacksmith's son in Oxfordshire to his geological work. Smith began as an assistant to a land surveyor and moved into mine-related projects, including excavations for canal-building. During the course of one such project he realised the significance of strata within layers of rock, and in subsequent surveys he could locate deposits of coal, iron and other minerals. Smith suffered throughout his life from financial problems which frustrated publication of his works; his map was published in 1815, but further works were never completed. Towards the end of his life, however, he finally received the scientific recognition that was his due.

Cambridge University Press has long been a pioneer in the reissuing of out-of-print titles from its own backlist, producing digital reprints of books that are still sought after by scholars and students but could not be reprinted economically using traditional technology. The Cambridge Library Collection extends this activity to a wider range of books which are still of importance to researchers and professionals, either for the source material they contain, or as landmarks in the history of their academic discipline.

Drawing from the world-renowned collections in the Cambridge University Library, and guided by the advice of experts in each subject area, Cambridge University Press is using state-of-the-art scanning machines in its own Printing House to capture the content of each book selected for inclusion. The files are processed to give a consistently clear, crisp image, and the books finished to the high quality standard for which the Press is recognised around the world. The latest print-on-demand technology ensures that the books will remain available indefinitely, and that orders for single or multiple copies can quickly be supplied.

The Cambridge Library Collection will bring back to life books of enduring scholarly value (including out-of-copyright works originally issued by other publishers) across a wide range of disciplines in the humanities and social sciences and in science and technology.

Memoirs of
William Smith, LL.D.

*Author of the 'Map of the
Strata of England and Wales'*

JOHN PHILLIPS

CAMBRIDGE UNIVERSITY PRESS

Cambridge, New York, Melbourne, Madrid, Cape Town,
Singapore, São Paolo, Delhi, Tokyo, Mexico City

Published in the United States of America by Cambridge University Press, New York

www.cambridge.org
Information on this title: www.cambridge.org/9781108245517

© in this compilation Cambridge University Press 2011

This edition first published 1844
This digitally printed version 2011

ISBN 978-1-108-24551-7 Paperback

Fourau, pinx.ᵗ   Ackermann & Cº London _ 1837.   T.A Dean. sculp.ᵗ

Wᵐ Smith LL.D.
aged 69.

MEMOIRS

OF

WILLIAM SMITH, LL.D.,

AUTHOR OF THE

"MAP OF THE STRATA OF ENGLAND AND WALES,"

BY

HIS NEPHEW AND PUPIL,

JOHN PHILLIPS, F.R.S., F.G.S.,

PROFESSOR OF GEOLOGY AND MINERALOGY IN THE UNIVERSITY OF
DUBLIN,

AUTHOR OF " ILLUSTRATIONS OF THE GEOLOGY OF YORKSHIRE."

Quæ fecimus, ipsi ...... ea nostra.

LONDON:

JOHN MURRAY, ALBEMARLE STREET.

1844.

" Ce que les minéralogistes les plus distingués ont fait dans une petite partie de
Allemagne, en un demi-siècle, un seul homme (M. William Smith, ingénieur des
mines) l'a entrepris et effectué pour toute l'Angleterre ; et son travail, aussi beau
par son résultat, qu'il est étonnant par son étendue, a fait conclure que *l'Angleterre
est régulièrement divisée en couches, que l'ordre de leur superposition n'est jamais
interverti ; et que ce sont exactement des fossiles semblables qu'on trouve dans
toutes les parties de la même couche et à des grandes distances.*

" Tout en payant an travail de M. Smith le tribut d'admirátion qui lui est dû, il
me sera permis de désirer que des observations ultérieures en confirment l'exactitude, et déjà, sur plusieurs points, les travaux des minéralogistes Anglais l'ont
confirmée."—D'Aubuisson, as cited by the Rev. W. D. Conybeare, Introduction
to Conybeare and Phillips's Outlines of the Geology of England and Wales, p. xlvi.

PRINTED BY RICHARD AND JOHN E. TAYLOR,
RED LION COURT, FLEET STREET.

THESE

MEMOIRS OF WILLIAM SMITH

ARE GRATEFULLY DEDICATED

TO THE MEMORY OF

THE LATE REV. BENJAMIN RICHARDSON,

OF FARLEIGH CASTLE, IN SOMERSET,

THE LOVED ASSOCIATE OF HIS EARLY STUDIES;

TO

THE GEOLOGICAL SOCIETY OF LONDON,

THE JUDGES AND REWARDERS OF HIS MERIT;

AND TO

Sir JOHN V. B. JOHNSTONE, Bart.,

OF HACKNESS, IN YORKSHIRE,

THE FIRMEST FRIEND OF HIS DECLINING YEARS.

PREFACE.

————◆————

THE following Memoirs are intended to illustrate the
life and character of the individual who has been
styled by competent authority the " Father of English Geology." They are drawn from authentic materials principally in the possession of the compiler,
who, after witnessing the workings of Mr. Smith's
mind and the changes of his fortune during the last
five-and-twenty years of his life, was called upon to
perform the duty of examining his voluminous unpublished papers. From these documents, all carefully dated, and the recollection of many graphic
stories in which Mr. Smith embodied his earlier
thoughts and struggles, the progress of his discoveries and some of the peculiarities of his private
history may be faithfully collected. No one interested in the annals of science would desire that
such records of one of its eminent cultivators should
be lost; but the writer—an orphan, who benefited by
his goodness—a pupil, who was trained up under his

care—feels it a privilege and a duty to endeavour to
save from neglect the memory of such a man.

Had Mr. Smith's published works been of such a
nature as to convey an adequate notion of the varied
operations of his mind, this attempt might have been
unnecessary, for the outlines of his scientific labours
have been already sketched in a few pages, but by
powerful hands.

The early progress of English Geology has been
admirably traced, in a wide and comprehensive view,
from a high point of knowledge, by Dr. Fitton*, who,
while determining by a just analysis the true and great
merits of William Smith, has happily restored to their
due honours the names of still earlier if less successful inquirers. When Professor Sedgwick hailed
Mr. Smith as a " great original discoverer" and the
" Father of English Geology," and delivered to him
the first Wollaston medal, he doubled the value of
this award by an eloquent address from the chair of
the Geological Society, embodying additional facts
and interesting original documents†. Dr. Buckland‡
and other eminent writers have done similar honour
to his memory. But these notices of the Author of
the Map of the Strata of England and Wales were
for the most part published in his lifetime, while yet
there was hope that some further if not more worthy
specimens of his geological studies than had appeared

* Notes on the History of English Geology, Phil. Mag. and
Annals, 1832. Reprinted from the Edinburgh Review, Feb. 1818.

† Address to the Geological Society, Feb. 18, 1831, on announcing the first award of the Wollaston medal.

‡ Address to the Geological Society, 1840.

might be given to the world; while yet there were feelings to be respected which forbad us to estimate the merits of the friend who still walked among us, as if they belonged to the province of history.

The events of Mr. Smith's life, and the progress of his scientific discoveries, are faithful and not uninteresting exponents of the times in which he lived; it was to sagacity and habits of observation operating on the circumstances in which he was placed, and not to the accidental concurrence of extraordinary phænomena with surpassing stretch of intellect, that his discoveries are due. They are but the effect of combining ideas which none had combined before, but which certainly could not have remained disunited many years later than the date of Mr. Smith's labours. A just account of these labours is therefore essential to the history of English Geology; and the present work, so far as it fulfils the condition of impartial justice, may be accepted as a contribution to that history, which can be offered in no other name than that of William Smith.

Another powerful motive for the present undertaking was supplied by the perusal of the unpublished, though not wholly unarranged, documents left by Mr. Smith. In his later years he had leisure to survey the laborious and painful course by which he had reached the temple of Fame, and approached the haven of repose; and feeling that in the early and critical periods of his mental progress, as well as in his declining years, he had been blessed beyond other men with the sympathy and support of faithful friends, he resolved to place on record his grateful sense of these favours. He was actively employed in this laudable purpose during the principal portion of the last year of his life, verifying from original documents, or recalling with the admirable exactness of a cultivated memory, the dates, the places, the persons, belonging to every remarkable scene in a varied life of seventy years.

On a review of other and earlier MSS., it was found that for most of the events which had been traced from memory in these latest compositions, original documentary evidence remained; and that from the same sources authentic proof could be brought of the performance of many kind and even noble acts, in favour of Mr. Smith and for the public good, by persons whose names are dear to science and their country. Records so honourable to those persons, which had been treasured with such care by the subject of their regard, have not been omitted in this summary of his life; but the Editor has purposely softened the darkest outlines of Mr. Smith's private and personal fortunes, because, though to others they seemed very melancholy, by himself they were only regretted as impediments in his adventurous enterprise.

York, Jan. 1, 1844.

LIST OF PLATES AND WOODCUTS.

MEMOIRS

OF

WILLIAM SMITH, LL.D.

———◆———

THE ancestors of William Smith were a race of farmers
who owned small tracts of land, and had been settled in
Oxfordshire and Gloucestershire for many generations.
In 1730, "William Smith of Sarsden, yeoman, eldest son
and heir-apparent of William Smith the elder, yeoman, of
Churchill," married Lucy Raleigh *, a daughter of Henry
Raleigh, yeoman, of Foscott in Oxfordshire. She had many
sisters (twelve or more) and one brother, who was accident-
ally killed. Her husband received as a marriage-portion
the sum of £110, in consideration of which his father settled on the bride, as a jointure, " one half-yard and half-
quarter of a yard land of arable, meadow and pasture land,
situate in Churchill field†."

* The signatures by Henry and Lucy Raleigh are both thus written, but
in the body of the deed the engrosser has written Rawleigh. From childhood William Smith was impressed by the belief that the Raleighs of Foscott were an obscure or forgotten branch of the ill-traced descendants of
Sir Walter Raleigh. (See Appendix A.)

† At this time and until 1787, a great part of the land in the parish of
Churchill was in " open field," that is, unclosed, but with the several
properties divided by well-recognised boundary lines, and known by the
descriptive term "Yard Lands." The labouring poor were often taken
into employment by the farmers in rotation, according to their " yard
lands." The land here described as five-eighths of a yard was at the enclosure consolidated to nearly ten acres, and was sold in 1807 for £700.

B

From these parents, who died in advanced age, sprung a third William Smith, who died in 1805, unmarried, and John Smith, who married Anne Smith of Longcompton in Gloucestershire, and died in 1777. The fruits of this marriage were a fourth William, the subject of this memoir, with two younger brothers and one sister. Anne Smith married again and died in 1807.

William Smith was born at Churchill, a village in Oxfordshire, on the 23rd of March, 1769, the year which gave birth to Cuvier. Of his parents he always spoke with great regard, but there is little in the recollections which he has preserved of them to show in what degree they contributed to form his remarkable character. His father he described as " a very ingenious mechanic," and mentions as the cause of his death a severe cold caught while engaged in the erection of some machinery. Deprived of this parent before he was eight years old, it was fortunate for him that his mother was a woman of ability, of gentle and charitable disposition, and attentive to the education of her children. An expressive pencil-sketch and a characteristic description, both from memory, record his devotion to his mother.

According to his own account, however, not only were the means of his instruction at the village school * very limited, but these were in some degree interfered with by his own wandering and musing habits. The rural games† in those " merrie daies " of England might sometimes attract the wayward and comparatively unrestrained scholar from his books; but he was more frequently learning of another mistress, and forming for after-life a habit of close and curious contemplation of nature.

After his father's death and his mother's second marriage, the person to whom he was principally to look up for protection was his father's eldest brother, to a portion of whose property he was heir. From this kinsman, who was but

* Founded by one of the former owners of Churchill and Sarsden.
† One of the merrymakings of Oxfordshire took place at Whitsuntide, under the denomination of "Whitsun-ale." The expenses of this festivity were not trifling. In 1720 and 1721 William Smith of Churchill was trea-

little pleased with his nephew's love of collecting the "pundibs*" and "poundstones," or "quoitstones†," and had no sympathy with his fancies of carving "sun-dials" on the soft brown "oven-stone‡" of the neighbourhood, he with great difficulty wrung, by repeated entreaty, money for the purchase of a few books fit to instruct a boy in the rudiments of geometry and surveying. But the practical farmer was better satisfied when the youth manifested an intelligent interest in the processes of draining and improving land; and there is no doubt that young William profited in after-life by the experience, if it may be so called, which he gathered in his boyhood while accompanying his relative (" old William") over his lands at Over Norton. Whatever he saw, was remembered for ever. To the latest hours of life he retained a clear and complete recollection of almost every event of his boyhood, and often interested young and old by his vivid pictures of what he had seen when a child. These notices would be swelled to an unreasonable degree by introducing the pleasant stories of

surer for the "Whitsun-ale," and in that capacity presented a complete account of the expenditure.

	£	s.	d.
The whole receipt in 1721 was	58	19	0
The whole expenditure	54	5	11

In the disbursements we find, properly vouched,—

	£	s.	d.
For Ribbands	11	19	0
... Malt	10	5	0
... Cake	5	3	0
... Excise	4	7	11
... the Fool	1	0	0
... the Fiddler	0	10	6
... the Morris	0	6	0
... the Lord	0	2	6
... the Lady	0	2	6
The Lord and Lady gave the Fool	0	1	6
For Bells	0	1	6

The Lord's man, the Lady's man and five maids, received nothing.

* Terebratulæ.

† A large Echinite (*Clypeus sinuatus* of Leske), not unfrequently employed as a "pound-weight" by the dairywomen.

‡ Named from its frequent use in the construction of ovens.

" the narrative old man;" but the following recollections,
written in his seventieth year, of events which had passed
fifty-six years before, are worth preserving as evidence of
this peculiar circumstantiality of memory.

"I was early a tall and strong-grown boy, and in my way
to London between twelve and thirteen years of age particularly noticed the great work of cutting down the chalk
hill at Henley-upon-Thames, and how the loaded carriages,
on an inclined plane, were made to bring up the empty ones.

"I was in London shortly after the riots of Lord George
Gordon; and at the time when the news of Rodney's defeat
of the French fleet arrived.

"There was then a half-penny toll for foot-persons passing Blackfriars Bridge; the Albion Mills (worked by steam
power) had just before been burnt down.

"Criminals were hanged at Tyburn, where there were cow-
houses with wood seats on top for persons to see the executions.

"From Manchester-square to the Edgeware-road and
Paddington, there were foot-paths entirely across open
fields. The buildings on that side of the square were un-
finished: but, as more connected with what relates to the
earth, I saw how the ground was made in Manchester-
square, for a poor fellow in turning his cart-load of slush
had let his horse and cart slip down, so that he was up to
his middle in mud endeavouring to extricate his horse just
as I passed by. This was on the east side of the square."

In 1783, and from this time to 1787, the young man,
without instruction or sympathy, prosecuted irregularly, but
with ardour and success, the studies to which his mind
was awakened. He began to draw, attempted to colour,
became tolerably versed in the geometry and calculations
then thought sufficient for engineers and surveyors, and by
these acquirements, at the age of eighteen, so strongly recommended himself to Mr. Edward Webb of Stow-on-the-
Wold, who had been appointed to make a complete survey
of the parish of Churchill, for the purpose of enclosure, that
he became assistant to that most able and excellent man,
and was taken into his family.

This was the critical moment; from this event flowed all
the current of his useful life, and to the same origin may be
ascribed many of the peculiar habits and feelings, the contrasted lights and shades, which diversified the character of
William Smith.

Edward Webb was, like his pupil, self-taught, and very
slightly acquainted with languages and general literature,
but possessed of great ingenuity and skill in mechanics,
mensuration, logarithms, algebra, and fluxions. His practice, as a surveyor, included many things now conceded to
the engineer, such as the determination of the forces of
water, and planning machinery*. His instruments were
commonly invented, often made and divided by himself;
peculiar pentagraphs, theodolites, scales, and even com-
passes and field books, of new construction, enriched the
office at Stow, and stimulated to thought and exertion the
young men who were fortunate enough to be placed in it†.

" I admired," says the subject of this memoir, " the talent
of my master, his placid and ever unruffled temper, and his
willingness to let me get on, for I required no teaching."

Speedily entrusted with the management of all the ordinary business of a surveyor, Mr. Smith traversed in continual activity the oolitic lands of Oxfordshire and Gloucestershire, the lias clays and red marls of Warwickshire;
visited (1788) the Salperton tunnel on the Thames and
Severn Canal, and (1790) examined the soils and circumstances connected with a " boring for coal in the New
Forest, opposite the Shoe alehouse at Plaitford." (MS.)
All the varieties of soil in so many surveys in different
districts were particularly noticed, and compared with the

* Among the recollections of the office, Mr. Smith used to quote the
following lines regarding the inventor of the slide-rule and steam-boats :—
" Jonathan Hull, with his paper skull,
He tried to make a machine;
But he, like an ass, could not bring it to pass,
And now he's ashamed to be seen."
The unfortunate boat was launched on the Warwickshire Avon.

† Mr. Richard C. Taylor, the eminent surveyor and geologist, now resident in the United States, was one of the pupils of Edward Webb.

general aspect and character of the country, the agricultural and commercial appropriations. The arrangement of the lias limestone beds in Warwickshire, contrasted with the neighbouring red marls at Inkborough, the boring for coal in some of the dark lias clays on the road to Warwick, the absence of arenaceous beds from the limestones of Churchill,—these were some of the points treasured in a mind capable of combining them at a future time.

That time arrived in 1791, when Mr. Webb transferred to his young friend the survey of an estate at Stowey in Somersetshire. Mr. Smith walked by Burford, Cirencester, Tetbury, Bath, Radstock, Old Down, Stoneaston, and Temple Cloud, to Stowey. Here he was surprised to find, as well as at High Littleton, the red marl, evidently similar to that of Worcestershire, similarly posited in regard to the lias and superincumbent rocks, and similarly employed for marling the land.

In his own words we read, " Coal was worked at High Littleton beneath the ' red earth,' and I was desired to investigate the collieries and state the particulars to my employer. My subterraneous survey of these coal veins, with sections which I drew of the strata sunk through in the pits, confirmed my notions of some regularity in their formation; but the colliers would not allow of any regularity in the matter of the hills above the ' red earth,' which they were in the habit of sinking through ; but on this subject I began to think for myself." The minute survey which Mr. Smith made of the High Littleton Collieries was continued at intervals through the years 1792 and 1793, and among the papers remaining, which demonstrate a perfect acquaintance with the effect of the faults, on the outcrops and depths of the coal, are an " Original Sketch and Observations of my first Subterranean Survey of Mearn's Colliery in the parish of High Littleton," and the following curious memoranda, dated June 15, 1793 :—

" Proposals toward making a model of the strata of earth, &c. in a coal country.

" Make a model of the strata of earth and coal at High

Littleton of the same materials of which they are composed, reduced to scale and placed in the same order in which they are found in sinking of pits; make a section of it.— N.B. Red ground and other soft materials may be mixed up with gum-water or some kind of glutinous substance.

" A model of the strata at High Littleton, Somerset, made of the same materials of which each stratum is composed, arranged in the same order as nature has placed them, and divided into sections, that may be taken apart to explain the method of mining for coal.

" The grays may be in pieces stuck together with gum, which will represent the water found in the joints.

" The brooks above ground may be filled with gum, which will be a good representation of water."

To this is added a comparison of scales most adapted for the representation intended, showing clearly the intention of the inventor, to make the model *true* and *proportionate* in vertical and horizontal measure.

The ability and perseverance manifested by Mr. Smith in his occupations near High Littleton, induced some of the gentlemen resident in the neighbourhood (particularly Mr. Mogg and Mr. Stephens of Camerton) to interest themselves in his professional success.

The occasion was favourable. At this era the services of civil engineers were much in request, and the principal duties entrusted to them were such as Mr. Smith was well qualified to perform. At the present time the accomplishments of an engineer are measured by a higher mathematical standard than in the times of Smeaton and Whitworth, because the practical problems committed to them for solution involve more varied dynamical and statical considerations than the cutting of a canal, or even the development of the powers of water and wind; but at all times the most valuable basis of good engineering practice must be a habit of accurate observation and a collectedness of mind, which meets unexpected emergencies with prompt and adequate combinations, appearing the result of intuitive sagacity rather than of treasured experience. That in these qualities, and a certain inventiveness, put to frequent proof, few ever surpassed Mr. Smith, will appear in the following pages. At the period we are speaking of, projects for new canals, accompanied by new schemes for conquering inequalities of level,—inclined planes, caissons, and locks of various kinds,— were actively canvassed, and Mr. Smith seized the occasion to procure instruments, extend his reading, and otherwise qualify himself as a competent candidate for the professional honours which seemed within his reach. This was a tide in his affairs, which, had he followed the middle current without stopping to examine the banks, would have led him on to fortune; and even under this great disadvantage of being subject to a strong deflecting force, his career was not unprosperous, and he joined with tolerable compactness the decisions of an engineer to the inquiries of a geologist.

In 1793 we find him engaged in executing surveys and complete systems of levelling for the line of a proposed canal. In the course of the operations which he performed in the summer and autumn, a speculation which had come into his mind regarding a general law affecting the strata of the district, was submitted to proof and confirmed. He had *supposed* that the strata lying above the coal were not laid horizontally, but inclined; that they were all inclined in one direction, viz. to the eastward, so as to successively terminate at the surface, and thus to "resemble, on a large scale, the ordinary appearance of superposed slices of bread and butter." This supposition was now proved to be correct by the levelling processes executed in two parallel valleys, for in each of the levelled lines the strata of " red ground," " lias," and " freestone" (afterwards called " oolite"), came down in an eastern direction and sunk below the level, and yielded place to the next in succession.

But at the same time it was known to Mr. Smith that the position of the strata of coal in Somersetshire was not generally conformed to that of the " red earth," " lias," and other beds above; the same thing was proved to him by an inspection of the colliery at Pucklechurch in Gloucestershire; he knew, besides, that the great faults which divide *all* the coal strata under ground were in general found not to divide *any* of the superincumbent rocks which formed the surface.

Geologists, who, at the present time, notwithstanding the devoted attention which has been paid to the phænomena of local displacement, find a clear conception of the causes at least difficult of attainment, may imagine the perplexity in the mind of a *discoverer* enlarged by the notion of general laws, but limited in the proof of of them to examples of local, and these partly exceptional phænomena. Mr. Smith felt this perplexity severely, but not long. The Canal Bill on which he was engaged received the sanction of Parliament in 1794, and one of the first steps taken by the judicious committee of management was to depute two members of their body to accompany Mr. Smith, " their engineer," on a tour of inquiry and observation regarding the construction, management, and trade of other navigations in England and Wales.

The tour extended altogether 900 miles, and occupied between one and two months; by one route the party reached Newcastle, and by another returned through Shropshire and Wales to Bath. Mr. Palmer and Mr. Perkins were gentlemen well acquainted with coal-working, and they willingly stayed to inspect every new invention applied to canals and collieries; but Mr. Smith's treasured object of consideration on the road, that which occupied all his thoughts in the interval of professional inquiries, was the aspect and structure of the country passed through, in order to determine if his preconceived generalizations of a settled order of succession, continuity of range at the surface, and general declination eastward, were true on a large scale.

It is needless now to say that his general views were justified; he found the strata from the vicinity of Bath and Bristol prolonged into the North of England, in the same general order of succession with the same general eastward dip. There is, however, one part of the conclusions adopted on this rapid survey from a postchaise which merits particular attention. He passed through York on the high road to Newcastle, and from that line, distant from five to fifteen miles from the hills of chalk and oolite on the east, he was satisfied of their nature by their *contours* and *relative position*, and their ranges on the surface in relation to the lias and " red ground" occasionally seen on the road. This was in fact the only authority he could rely upon for drawing, in 1800, the continuations of the chalk of Wiltshire and the oolite of Somersetshire through the eastern parts of Yorkshire, but he drew them with a considerable approximation to accuracy. The following notice of this tour was written in June 1839, nearly in the same words he had often used before in narrating it :—

" After the passing of the Act of Parliament for the canal in the summer of 1794, and some preliminary business, it was determined at one of the meetings that, as canals were not then known in those parts, and before the works should be commenced, two of the committee, Dr. Perkins, Samborn Palmer, Esq., and myself, should make a tour throughout England and Wales to procure the best information we could on canals, and report the same to the company of proprietors.

" This was joyous intelligence to me. I wished to travel; for I foresaw that the truth and practicability of my system must be tested far and wide before its uses could be generally known and its worth duly appreciated.

" I thought, of course, no one could do this so well as myself; and the result of my observations on this tour may be considered as the first part of *the interminable labour of working out the truths of the science*; for it plainly appeared that it was to become a system of experimental philosophy, which would embrace the whole surface of the globe.

" No journey purposely contrived could have better answered my purpose.

" To sit forward in the chaise was a favour readily granted; my eager eyes were never idle a moment; and post-haste travelling only put me upon new resources. General views, under existing circumstances, were the best that could have been taken, and the facility of knowing, by

contours and other features, what might be the kind of
stratification in the hills, is a proof of early advancement in
the generalization of phænomena.

" In the more confined views, where the roads commonly
climb to the summits, as in our start from Bath to Tetbury,
by Swanswick, the slow driving up the steep hills afforded
me distinct views of the nature of the rocks ; rushy pastures
on the slopes of the hills, the rivulets, and kind of trees, all
aided in defining the intermediate clays; and while occasionally walking to see bridges, locks, and other works on the
lines of canal, more particular observations could be made.
Much, however well observed but depending upon memory, would of course be lost, for this was all foreign to
the purport of our journey; and also another important inquiry on *coal* and *collieries*, for which we had each, by
agreement, provided an extra memorandum-book.

" My friends being both concerned in working coal, and
Mr. Palmer having it in his own estate, they were interested
in two objects; but I had three, and the most important
one to me I pursued unknown to them; though I was continually talking about the rocks and other strata, they
seemed not desirous of knowing the guiding principles or
objects of those remarks; and it might have been from the
many hints perhaps mainly on this subject, which I made in
the course of the journey, that Mr. Palmer jocosely recommended me to write a book of hints.

" We had nothing official to notice until we arrived at the
Thames and Severn Canal at Thames head. The tunnel,
however, 4¼ miles long, through Salperton Hill, and 90 or
95 yards beneath the surface, was a main object; and that
on the Worcester and Birmingham Canal at Kingsnorth,
then making, was another. The former was through the
summit edge of the stonebrash range of the Cotteswold Hills,
many years before well known to me ; and the latter through
the red marl and red sandstone, well known at High Littleton and other Somersetshire collieries to be an *unconformable cover to the coal-measures.*

" Saw the same red rock again at the tier of locks on the

high side of Birmingham, and at various places on the road to Derby. Saw the silk mills. Went to Derby and Ripley Collieries, where I was surprised to see the long, straight-grained pieces of coal piled up and sold in stacks four feet square.

" Went to see Chatsworth, where I remember the varie-gated columns of millstone grit,—the bold hills of that rock thereabout, and the limestone rocks and tufa of which the inn was built at Matlock.

" In staying hereabout two days we met with Benjamin Outram, the engineer of the Cromford Canal, who took us into the tunnel at Butterley Park, then making : not a stone of the great ironworks since established was then laid.

" Chesterfield, in the midst of a rich coal-field, was then a poor place, and is so now.

" We found the small steam-engines much better applied to raising coal in Yorkshire than in Somersetshire, where not more than one (I believe), badly constructed, was then in use.

" Flat ropes were in use, and at Hisley Wood or White-lane Colliery, on Earl Fitzwilliam's estate, I stood on one end of a cross-bar to which corves were suspended, and Mr. Perkins on the other, and we were very smoothly let down a little more than mid-depth of the pit to see Mr. Cur's, so called, sliding-rods ; when, on being stationary and directed to look up, we saw the ascending corves over our head without knowing that they had passed us at the mid-depth enlargement of the pit.

" On reversing the motion of the engine we were soon again on the surface, and simply and easily as this seemed to be effected, by grooves down from top to bottom of the pit, for opposite ends of the cross-bars to move in with an en-larged place mid-way for their passing, my learned friend could not understand the mechanism, though I was occupied nearly the whole of a long stage in explaining it to him.

" The rocks of the Yorkshire coal-field, everywhere so well developed, opened to my mind new views of the facility of obtaining on the surface clear notions of the coal-measure stratification; and at Banktop, near Barnsley, I found some of the rocks of this series strongly developed.

" Leeds being near the extent of the coal-field, we found that further north there were no canals, but determined on seeing York Minster; and thus, in crossing Tadcaster Moor, I had a clear view of the magnesian limestone, which is a rock unknown in the south.

" From the top of York Minster I could see that the Wolds contained chalk by their contour.

" We here had time enough to indulge ourselves with a good dinner and a pine-apple at the Black Swan, and re-solved upon a run up to Newcastle, to see the celebrated col-lieries there; and after the first stage from York, I recognized in the Hambleton Hills the features of the Cotteswold Hills viewed from the Vale of Gloucester; saw near Thirsk the red marl in the road, and found that along Leaming-lane we were travelling upon red sandstone. The yellow lime-stone appeared again at Peirce Bridge, and at Ferry Hill they were working coal under it.

" Here it presented a well-defined escarpment boundary to the Durham coal-field, as it did to that of Yorkshire; but these northern coal-measures were observed to be much more obscured by a thick cover of loose and mixed matter.

" We arrived at Newcastle on Saturday afternoon, time enough to get to Heaton Colliery, but unfortunately too late for me to go down in the pit; but a very intelligent overlooker kindly drew with his stick on the dust a plan of the mode of working the coal, which to me was perfectly intelligible.

"The railways to the staiths on Tyneside were then mostly of wood, or wood plated with iron; and such was the state of machinery, that at Heaton Colliery the deep water was raised by a steam-engine into a pool on the surface, and at other times in the twenty-four hours from the pool, by much larger pumps, to the top of two high water-wheels, which raised the coal.

" We did not expect to see things so managed in the North; and I was surprised to see the fires they kept, and

other contrivances for promoting ventilation, as in the
Somersetshire collieries there is no want of a good current
of pure air.

" I had observed that my friend Palmer's string of questions sometimes produced a shyness in obtaining answers,
and therefore I used to proceed upon the principle of give
and take; and in thus offering my exchange of knowledge
of the mode of working coal in Somersetshire, 1000 yards
down the steep slope of 1 in 4, and perfectly dry and in
good air, 100 to 250 perpendicular yards beneath the bottom of the pumps, I believe the honest manager of Heaton
Colliery thought I was telling him a travelling story.

" The mode of dividing their shafts and mother-gates by
brattices of wood-work seemed inconvenient and unphilo-
sophical, and we, rather dissatisfied, hastened back through
Ripon and Harrowgate, where the M.D. took a nauseous
draught of sulphur-water as we sat in the chaise.

" We had crossed the yellow limestone between Ripon
and Ripley."

Engaged for six years in setting out and superintending
the works on the Somersetshire Coal Canal, Mr. Smith
found but few opportunities of making known to scientific
persons the peculiar generalizations which had taken possession of his mind. But in the execution of these wishes
he was putting them daily into practice, informing the contractors what would be the nature of the ground to be cut
through, what parts of the canal would require unusual
care to be kept water-tight, what was the most advantageous
system of work. Another singular advantage attended this
engagement; the notions which up to this time he had obtained regarding the distribution of organic remains were
comparatively vague ; he found peculiar plants in the " clift"
above the coal, particular shells in the lias and oolites, but
none in the red ground, and had combined these simple
facts so far as to see that " each stratum had been successively the bed of the sea, and contained in it the mineralized
monuments of the races of organic beings then in existence."
But it was the necessity of a close and accurate knowledge

of the different sorts of rock, sand, and clay, which were to
be cut through on the line of the canal, which led him to
examine minutely and scrupulously into the distribution of
the " extraneous fossils" which he had been in the habit of
collecting. The result was a proposition which he proved
to be locally true, and of practical value, " that each stratum
contained organized fossils peculiar to itself, and might, in
cases otherwise doubtful, be recognized and discriminated
from others like it, but in a different part of the series, by
examination of them." He now remarked also the contrast
between the rounded state and mixed condition of the fossils which lay in gravel deposits, and the sharply preserved
specimens lying in natural associations in the strata; and
thus acquired a notion of the distinction between what were
afterwards named diluvial and stratified deposits.

The possessor of all these generalizations, now (1795)
twenty-six years of age, was still shrouded in the obscure
village of High Littleton, but in this year he removed to
Bath, and took up his abode in the central house of a short
range of buildings called the Cottage Crescent, which occupied a picturesque and elevated site south of that city.
" From this point," says he, " the eye roved anxiously over
the interesting expanse which extended before me to the
Sugar-loaf mountain in Monmouthshire, and embraced all
in the vicinities of Bath and Bristol; then did a thousand
thoughts occur to me respecting the geology of that and
adjacent districts continually under my eye, which have
never been reduced to writing." He continued to direct all
the operations on the Somerset Coal Canal, and very copious
note-books attest the constancy and exactitude of his attention to that occupation. To this cause, indeed, may be
ascribed the extreme rarity of any essays or even memoranda from which the progress of his geological studies can
be gathered.

That in Jan. 1796, he had begun to commit his thoughts
to paper in a lucid arrangement for publication, the written
proofs remain; in 1797, he drew a larger general plan for
such a work; but not till 1799, after his engagement ceased

with the Coal Canal Company, did he make public his intention to compose a general work on the Stratification of Britain, or enter on the prosecution of an actual survey of the geological structure of the whole of England and Wales.

In the execution of the canal, Mr. Smith had found the means of applying his newly-acquired knowledge to useful practical problems,—such as how to draw the line through a country full of porous rocks, so as best to retain the limited supplies of water which frequent mills left to the navigation—where to place bridges on a good foundation—how to intercept and conduct the springs, and where to open quarries of proper stone. We find him also engaged, as early as 1796, in the short intervals which could be snatched from the main business before him, in putting to practical proof his theoretical views of the earth's structure, and the properties of the mixed calcareous and argillaceous strata in the hills near Bath, by a new and successful process of land-draining.

Under these circumstances, it was impossible for a plain, simple-minded and enthusiastic man, to avoid explaining his views to such intelligent persons as would listen to them ; but Mr. Smith found few auditors who interested themselves in his speculations, any further than as they appeared to have immediate practical results in agriculture or mining. The very intelligent land-steward of the Marquis of Bath, Mr. Thomas Davis, (author of the excellent Report on Wiltshire, presented to the Board of Agriculture in 1794,) when informed of the constitution of the Wiltshire hills and vales, and the relation they thus held to neighbouring tracts, was chiefly moved by the obvious light such discoveries shed on the agricultural appropriation of soils, and remarked, " that is the only way to know the true value of land."

Even such sympathy was highly prized by the modest " Father of English Geology," who, in his latest years, when geology had claimed in a high degree the public favour, frequently recounted, among many instances of mortifying disregard which he had experienced, this apparently slight and solitary sentence of encouragement. It is to be regretted, that of that period in Mr. Smith's mental progress when the grand ideas of a new science were struggling for distinctness and generality, so few written monuments, and those not the earliest, remain. The essays of this period which have been found indicate the existence of more and earlier efforts, and, being all marked with the place where and the date when they were written, possess a peculiar interest and authenticity. The documents of this nature, belonging to the years 1796, 1797, and 1798, which will be referred to in the following pages, appear to have been written in the short and detached intervals of leisure left by almost constant business, *not at home*, but at various points to which the daily occupation of a canal engineer conducted him. It is perhaps by mere accident that any of these papers remain; for they have been all transcribed into "*a book*" which was intended for publication, but is probably not in existence, nor has any index of its contents been recovered.

The earliest connected remarks which have been found bear the date of January 1796, and relate to organic remains, and their distribution in the different strata. The vicinity of Bath is rich in fossils, and fine collections were formed there previous to Mr. Smith's researches: it might be after inspecting some of these treasures, whose full value was so entirely unknown to their owners, that the following reflections, which strikingly illustrate the enlarged state of his own views at that period, were penned.

"Dunkerton, Swan, Jan. 5, 1796.

" Fossils have been long studied as great curiosities, collected with great pains, treasured with great care and at a great expense, and shown and admired with as much pleasure as a child's rattle or a hobby-horse is shown and admired by himself and his playfellows, because it is pretty; and this has been done by *thousands who have never paid the least regard to that** wonderful order and regularity with which

* Underlined in the original.

c

18

Nature has disposed of these singular productions, and assigned to each class its peculiar stratum."

Gifted in a very uncommon degree with that philosophical faith in the generality and harmony of natural laws which is a characteristic of discoverers in natural science, Mr. Smith was at the same time remarkably disinclined to indulge in himself, or even to tolerate in others, mere speculations in geology. Whatever of this nature he found in the circle of his reading was severely judged, by a close collocation of the hypothesis which had been advanced with the phænomena of stratification which he had entirely established. These judgments might be erroneous in cases which required the knowledge of other data, not then collected, for a true and general solution; but the very unreasonableness of raising the standard of his own discoveries in a limited region, for condemning a speculation perhaps founded on other truths occurring elsewhere, shows how firmly these discoveries, and the inferences belonging to them, were established and fortified in his mind. The following passage, written in January 1796, might have been acknowledged by the author to contain his real opinions forty years later :—

" Therefore every man of prudence and observation who has paid the strictest attention to mineralogy, the structure of the earth, and the changes it has undergone, will be very cautious how he sets about to invent a system which nature cannot conform to without having recourse to subterraneous fires, volcanic eruptions, or uncommon convulsions, by which every hill and dale must have been formed, and every rock must have been rent to form those chasms, which, in comparison to the strata they are found in, are no more than sun-cracks in a clod of clay; yet such has been the language of ingenious men, who have set their theoretical worlds a-going without either tooth or pinion of nature's mechanism belonging to them."

In October and November of this year (1796), we find him returning to the contemplation of organic remains; discussing the circumstances which attend the sparry substance occupying the place of the shell, which has been removed, in the lias, and the empty cavity, where the shell was, surrounding a loose stony cast of the interior, in the freestone (oolite).

That his mind was now actively employed in tracing out the bearings of the extensive subject before him, will be evident from the following extract, dated August 1797 :—

"Locality of Plants, Insects, Birds, &c., arises from the nature of the strata.

"Where art has not diverted the order of things, and nature is left to herself, a considerable locality may be observed in many animals and vegetables, as well as mineral productions, by which they are evidently attached to particular soils to such a degree that, if this subject were studied with attention, it would form one of the principal external characteristics of the strata underneath. Though it may seem mysterious to some that birds, beasts, insects, &c., which have the liberty of roving at pleasure, should feel any particular attachment for this or that soil, yet the wonder ceases when we consider how the chain of natural things is linked together, and how these creatures are taught to cull their food from insects that are lodged in, or seeds that are produced from, particular plants that grow upon particular soils."

In the same year (November 3), we have a specimen of the way in which Mr. Smith was proceeding to record the localities of fossils.

" 1. Snakestone, 11 inches diameter.

" Found near the bottom of a rough bed of bastard freestone (inferior oolite), which lies upon a thick bed of sand and sand burs.

"*Remarks.*—The surface of this stone is covered with marks which have some resemblance to the leaves of plants [edges of the Septa], but on breaking any stone of this sort these marks are found to represent natural divisions of the fossil, which are so linked or dove-tailed one into another in this curious manner as not to be separated without violence after the joints are considerably loosened.

c 2

" 2. Part of a snakestone, about 4½ inches diameter.
" Found in the same bed about two miles distant from
the place where No. 1 was found.

" *Remarks.*—The loose joint at the end of this stone will
fully explain what I have said about the leaves and joints of·
No. 1.

" 3. Snakestone, about 3 inches diameter, of a bluish cast.
" Found in the blue lias limestone."

On this occasion we may remark the entire want of scientific or technical terms in all the statements relating to organic remains and the arrangement of the strata. The
names which Mr. Smith uses for designating the objects are
merely the provincial terms of the country where he was
located ; the new *ideas* which were connected with them in
his mind, were the fruit of his own observation and reasoning, and acquired new forms of expression at a later time.
The localities of all the specimens in Mr. Smith's possession
were carefully written *on the stones themselves*.

A manuscript, dated December 2, 1796, Dunkerton, Swan
Inn, headed "Strata in general, and their position," and
evidently intended for publication, commences thus :—

" The strata being found as regular on one side of a rivulet, river, deep valley or channel as on the other, over an
extent of many miles, when proper allowance is made for the
inclination and for the variation of the surface, is it not reasonable to suppose that the same strata may be found as
regular on one side of a sea or ocean as on opposite sides of
a deep valley upon land, and if so, and the continuation of
the strata is general, what is their general direction or drift?
Is it in straight lines from pole to pole, or in curved lines
surrounding the globe regularly inclined to the east ?"

After hinting at a general cause for such an assumed regularity, he adds, " But all theories are best built on practical rules, which will enable any one to make such observations for himself as must carry conviction along with them;
for a work so novel as this must expect to find some who
will hardly believe what is plain to be seen ; for all men do
not see alike, nor can patiently trudge through the dirt to

search for truth among the stubborn rocks where nature has best displayed her.......Shall, therefore, describe a number of quarries, cliffs, &c., at a great distance, &c. See Book——."

In what seems to be the continuation of this paper, we see the predominant desire of the author to establish the certainty and generality of the inclination of strata, which he had proved on a limited scale near Bath.

"If the strata lay horizontal, every part of the sea-shores would present the same beds at the water edge instead of that wonderful variety which is found on the coast and banks of every river and rivulet in the kingdom, especially those that run in an east and west direction, or nearly so. In such situations the young mineralogist may soon be convinced of that wonderful regularity which nature has adopted, especially if the shores are rocky; he will there find that, independent of partial and local dips which appear in different quarries of the same stone, the outlines or top and bottom layers of each complete stratum or class of stones or earth, considered as a mass, have a general tendency toward the eastern horizon."

This MS., revised and altered in the first lines, now stands, with several others written in 1797 and 1798, as part of the introduction to a work which is thus sketched in 1797.

" Plan of the work.

" To be divided into two parts.

" The first of which should treat of the strata of the earth in a general way.

" And the second enter into the particulars of each stratum, with the fossils and minerals that have hitherto been discovered, with their connexion and dependence one upon another. Though it is impossible for the labours of any one individual ever to accomplish a thousandth part of what is proposed by this section, yet when a regular system is established which has nature for its prototype, every one will be enabled to contribute his mite, and carry it on from time to time till after ages may get a tolerable description of the habitable world."

On the next page, dated Nov. 5, 1797, the details of this plan are considered thus:—

"Quere. The best method of explaining the order of the strata, whether by plain coloured maps, or varied black lines, or any variety of coloured lines.

"Sections of the strata in different directions will be necessary to show their various inclinations.

"In the general section each principal stratum should be numbered with progressive numbers, beginning at the eastern strata of the kingdom, or till that can be accurately ascertained, at some stratum that forms a grand feature therein, as for instance the chalk, which I would number 1; and those lesser strata which are contained within it or generally attached to it, or form any subdivisions therein, I would call 1 *a*, 1 *b*, 1 *c*, &c. If any stratum should be omitted, or a new one discovered, [it] may be brought into these numbers by marking it 1 *a a*, &c.

"After the general section of a country or district should follow a large section of each stratum, with its concomitant small strata on the left-hand page, and on the right-hand page* with drawings and descriptions of such peculiarities as the principal stratum, or those connected with it, are found to contain, whether the exuviæ of marine animals, vegetable impressions or fossil wood: coal and metals of every description.

"The same numbers may refer to an explanation of the chemical properties of each substance, so far as discovered; this may be placed at the end of the book, or make a separate volume, where these properties may be more minutely examined than can consistently be done in the body of the work, which is intended to form a true representation of the order of nature, with no more digressions from the main subject than are absolutely necessary to make it intelligible.

"All the plates should be bound at the end of each volume in a peculiar manner, with cuts or small plates of each shell, &c. amidst the letter-press; and where there is room in the plate the shells may be dotted out from and drawn

* The direction as to the pages has been crossed by the pen at a later date.

at the end of each stratum they are found in: these, as well as the strata, to make them more striking, should be coloured*."

In February 1798, we find as part of the Introduction to this contemplated work, an interesting notice of some of the steps by which the author was conducted to his general conclusions.

" It will be readily admitted by all classes of men, from the most accurate observers of nature to the simplest peasant, that there is some degree of regularity in the strata from whence our building materials are generally collected. Masons, miners and quarrymen can identify particular beds of stone dug many miles apart; indeed every cliff and quarry presents a true section of a great many beds of stone, which may be found of the same quality and in the same position in all or most of the neighbouring cliffs and quarries. And this regularity is nowhere more conspicuous than in the lias quarries of Somersetshire, from whence these observations first took their rise about seven years since.

" For the stratification of stone struck me, who had not been accustomed to such appearances, as something very uncommon, and till I had learned the technical terms of the strata, and made a subterraneous journey or two, could not conceive a clear idea of what seemed so familiar to the colliers; but when these difficulties were surmounted, and an intelligent bailiff accompanied [me], I was much pleased with my peregrinations below, and soon learnt enough of the order of the strata to describe on a plan the manner of working the coal in the lands I was then surveying.

" Being engaged soon after to survey the lands and take the levels of a canal that was proposed to be made from the collieries to Bath, I observed a variation of the strata on the same line of level, and soon found that the lias rock which about three miles back was full 300 feet above this line was now thirty feet below it, and became the bed of the river,

* These ideas were in great degree unexecuted till 1817. The method of notation here suggested for the strata, might be revived with advantage at the present day.

and in that direction did not appear any more at the surface. This induced me to note the inclination of the same rock, which I knew was to be found at the head of two other valleys lying each about a mile distant from, and in a parallel direction to, the one just described, and accordingly found it to dip the same to the south-east, and sink under the rivers in a similar manner.

" From this I began to consider that other strata might also have some general inclination as well as this (though I had been frequently told by the colliers that there was no regularity in the strata above ground), yet by tracing them through the country some miles, I found the inclination of every bed to be nearly the same as [that of] the lias; and notwithstanding the partial and local dips of many quarries which varied from this rule, I was thoroughly satisfied by these observations that everything had a general tendency to the south-east, and thence concluded there could be none of these beds to the north-west, the truth of which conjecture was soon verified by a tour of observation through the northern parts of this kingdom." [The journey here alluded to was performed in 1794. See p. 11.]

Fairly engaged in compiling an account of his discoveries, Mr. Smith naturally turned to inquire what had been done previously in this department of knowledge in his own country. His mind tending in a very strong degree to the path of observing and accumulating facts which might correct and strengthen the general views which he already possessed, he sought, in county histories and other probable sources, information such as he needed. Plot's Oxfordshire, Morton's Northamptonshire, and Woodward's Catalogue of Fossils, became his favourite study; and probably no other reader has drawn from those curious but neglected works so many valuable geological data: Morton's drawings of Ammonites gave him points of lias, Plot's Echinida fixed localities of inferior oolite; and thus by applying the principle of " identifying the strata by their imbedded organic fossils," he was rapidly enlarging his coloured sketches for a geological map of England and Wales.

Rev.ᵈ Benjamin Richardson

from a Silhouette in the possession of Mᴿˢ Richardson.

The most prevalent notion in the works which Mr. Smith could then consult, regarding the forms and localities of organic remains, was the vague and irrational belief (founded on a misconception of the meaning of Scripture, but handed down even to these days as if to demonstrate the indestructibility of a popular error), that these relics of more ancient systems of life were all buried in the solid strata of the earth by the operation of the general deluge. This notion appears never to have influenced for a moment the mind of Mr. Smith, who in his MS. of this period (1797–98), not only denies the unsatisfactory hypothesis, but places in direct contrast with it his own views, that it is the gravelly deposits scattered over the earth's *surface*, and containing *bones of quadrupeds* and *rolled and transported rock masses holding fossils which had been previously imbedded and petrified*, which should be ascribed to diluvial action.

" I verily believe that those waters did not penetrate to such a depth, nor disturb the strata so much as has been imagined ; yet the effects of a deluge are very visible upon the surface of the earth, and to a great depth beneath, especially in low lands by the sides of large rivers, where great quantities of gravel, sand, and mud, are generally collected, in which the remains of trees and animals are frequently found preserved entire. And these are the things which may be reckoned among the most perfect proofs of a deluge, but at the same time they must not be mistaken for or confounded with fossils of a very different description and of a different origin, such as the exuviæ of marine animals and vegetable impressions, which are always found regularly imbedded in the solid strata, and composed of the same matter as the mass, whatever it may be ; whereas the horns of the moose deer, and teeth of elephants, which are found in loose gravel or mud, are real substances which contain all the properties of horn or ivory, and none of those which belong to the stone or clay which surrounds them.

"But gravel-stones always contain fragments of such shells, or other marks, as are always found in the solid strata of which they originally formed a part, till torn from their native beds by the impetuosity of the retreating waters, which is a decided proof that such shells were formed prior to the deluge, as their masses are all rounded off by attrition in water, while those of the same sort which are still imbedded in the strata, remain perfect and entire."

In March 1798, Mr. Smith purchased a small but beautiful estate in a deep valley within three miles of Bath, almost overgrown with wild wood, hiding in its bosom a sheet of water and a small mill. Through this retired possession the canal was cut, without greatly injuring its remarkable beauty; and under Mr. Smith's fond and tasteful attention the scene was partly cleared, the pond expanded to a lake, the cottage became a comfortable home, in which he passed many happy and thoughtful hours. He did not, however, at any time reside long in this favourite retreat, but took up his station for about a year at the village of Mitford, near Bath, and engaged in the last duties which he performed as resident engineer to the Coal Canal.

Owing to a misunderstanding with the Company, this occupation ceased in June 1799, and Mr. Smith felt and acknowledged that a new æra in his life had arrived. He was not only at liberty, but placed under the necessity to consider the best means of making known his geological system, and of founding upon it a professional practice, which might provide the expense of travelling to verify and extend his knowledge, and fill up the outlines of a geological map of England and Wales.

In these objects, which were ever closely associated in his own mind, he was successful; the most valuable portions of his discoveries soon became public property, and he quickly acquired extensive employment in the practical applications of these discoveries to mineral surveying and draining of land on a large scale. The extensive diffusion of his fame and opinions, which now began, was owing to no actual and authorized publication, but to continual discussions and explorations with several active friends, oral communications and exhibitions of maps at agricultural meetings (then frequent), and circulation of

MS. copies of tabular expositions of the series of strata at that time determined.

His views at this epoch appear by the following notice:—

" During my five years' close confinement to practical engineering on the Coal Canal, my much-wished-for opportunity of collecting observations enough from the ranges of the different strata to make an accurate delineation of the stratification throughout England were suspended.

" I had seen enough by my tour of August 1794, to satisfy myself of the practicability of doing it, and often wasted much time in poring over maps, in contriving how the ranging edges and planes of the different strata could best be rendered intelligible: models were thought of, and one small map was cut along the edges of some of the strata with a view of defining their extent, and of showing how one stratum was successively covered by another.

"I drew in colours, on a map of the vicinity of Bath, and on Day and Masters' County Survey, all [that had been observed] very accurately to a certain extent, which embraced an interesting but intricate variety of strata in hills around Bath; and some small maps of England were spoiled by speculating on the ranges of stratification without sufficient data. The intricacies in their marginal edges were such that I found, to mark point by point, as the facts were ascertained, was the only way in which I could safely proceed.

" My experience in what I had done upon the Somersetshire map was sufficient to convince of this, and that to make a map of the strata on a scale as large as Cary's England (five miles to an inch) with sufficient accuracy, much of it should first be drawn on a larger scale."

It was fortunate for Mr. Smith and for the progress of his views, that he gained at this time the friendship of a man singularly competent to estimate the truth and value of these views, and both able and willing to advocate the merit of their author. The Rev. Benjamin Richardson was at this time living in Bath, and possessed a choice collection of local fossils, mostly gathered by his own diligent hands. Extensively versed in natural history, and generally well acquainted with the progress of science, he was perfectly enthusiastic in following out, and liberal in enabling others to prosecute, new and ingenious researches, especially if they tended to practical and public good. He knew accurately the country in which Mr. Smith had principally worked, and was acquainted with the views entertained on the subject of fossils, which had been recorded in books, or were adopted by the collectors, who were even then celebrated in the vicinity of Bath. He had no knowledge of the laws of stratification and the connexion between the forms of organic life and the order of superposition of the strata; while, on the other hand, his new friend had very little knowledge of the true nature of these organic forms, and their exact relation to analogous living types. The result of a meeting between two such reciprocally adjusted minds was an electric combination; the fossils which the one possessed were marshalled in the order of strata by the other, until all found their appropriate places, and the arrangement of the cabinet became a true copy of nature.

That such fossils had been found in such rocks was immediately acknowledged by Mr. Richardson to be true, though the connexion had not before presented itself to his mind; but when Mr. Smith added the assurance, that everywhere throughout this district, and to considerable distances around, it was a general law that the " same strata were found always in the same order of superposition and contained the same peculiar fossils," his friend was both astonished and incredulous. He immediately acceded to Mr. Smith's proposal for undertaking some field examinations to determine the truth of these assertions, and having interested in this object a new and learned associate, the Rev. Joseph Townsend (author of Travels in Spain), they at once executed the project. Among other places visited with this view was the detached hill on which Dundry Church is conspicuously elevated. From its form and position in respect of the lias of Keynsham, Mr. Smith had inferred that this hill was capped by the lowest of the Bath " freestones" (inferior oolite); and, from his general views, expected to find in that rock the fossils which it contained near Bath, on the westward rise, which he believed to affect all the strata near Bath above the coal. It is needless now to say, that examination confirmed both the inference of the character of the rock and the conformity of its organic contents. The effect of this and other illustrations of the reality of Mr. Smith's speculations was decisive. In general literature and especially in natural history, Mr. Smith was immeasurably surpassed by his friends, but they acknowledged that, from his labours in a different quarter, a new light had begun to manifest itself in the previously dark horizon of geology, and they set themselves earnestly to make way for its auspicious influence.

One day, after dining together at the house of the Rev. Joseph Townsend, it was proposed, by one of this triumvirate, that a tabular view of the main features of the subject, as it had been expounded by Mr. Smith, and verified and enriched by their joint labours, should be drawn up in writing. Richardson held the pen and wrote down, from Smith's dictation, the different strata according to their order of succession in descending order, commencing with the chalk, and numbered, in continuous series, down to the coal, below which the strata were not sufficiently determined, according to the scheme already noticed, p. 23.

To this description of the strata was added, in the proper places, a list of the most remarkable fossils which had been gathered in the several layers of rock. The names of these fossils were principally supplied by Mr. Richardson, and are such as were then, and for a long time afterwards, familiarly employed in the many collections near Bath. Of the document thus jointly arranged each person present took a copy, under no stipulation as to the use which should be made of it, and accordingly it was extensively distributed, and remained for a long period the type and authority for the descriptions and order of superposition of the strata near Bath. The following is a copy of the original document in Mr. Richardson's handwriting, presented to the Geological Society in 1831 :—

No. I.—Order of the STRATA and their imbedded ORGANIC REMAINS, in the vicinity of BATH; examined and proved prior to 1799.

Strata.	Thickness.	Springs.	Fossils, Petrifactions, &c. &c.	Descriptive Characters and Situations.
1. Chalk	300	Intermitting on the Downs	Echinites, pyrites, mytilites, dentalia, funnel-shaped corals and madrepores, nautilites, strombites, cochlæ, ostreæ, serpulæ	Strata of Silex, imbedded.
2. Sand	70			The fertile vales intersecting Salisbury Plain and the Downs.
3. Clay	30	Between the Black Dog and Berkeley.		
4. Sand and Stone	30	Hinton, Norton, Woolverton, Bradford Leigh.		Imbedded is a thin stratum of calcareous grit. The stones flat, smooth, and rounded at the edges.
5. Clay	15			
6. Forest Marble	10		A mass of anomie and high-waved cockles, with calcareous cement	The cover of the upper bed of freestone, or oolite.
7. Freestone	60		Scarcely any fossils besides the coral.	Oolite, resting on a thin bed of coral.—Prior Park, Southstoke, Twinny, Winsley, Farley Castle, Westwool, Berfield, Conkwell, Monkton Farley, Coldhorn, Marshfield, Coldaslston.
8. Blue Clay	6			
9. Yellow Clay	8	Above Bath.		
10. Fuller's Earth	6			Visible at a distance, by the slips on the declivities of the hills round Bath.
11. Bastard ditto, and Sundries	80		Striated cardia, mytilites, anomie, pundibs and duckmuscles.	
12. Freestone	30		Top-covering anomie with calcareous cement, strombites, ammonites, nautilites, cochlæ hippocephaloides, fibrous shell resembling amianth, cardia, prickly cockle, mytilites, lower stratum of coral, large scollop, nidus of the muscle with its cables	Lincombe, Devonshire Buildings, Englishcombe, Englishbatch, Wilmerton, Dunkerton, Coomhay, Monkton Coombe, Wellow, Mitford, Stoke, Freshford, Claverton, Bathford, Batheaston and Hampton, Charlcombe, Swanswick, Tadwick, Langridge.
13. Sand	30		Ammonites, belemnites	Sand hurs.
14. Marl Blue	40	Round Bath.	Pectenites, belemnites, gryphites, high-waved cockles	Ochre balls.—Mineral springs of Lincombe, Middle Hall, Cheltenham.
15. Lias Blue	25			
16. Ditto White	15		Same as the marl with nautilites, ammonites, dentalia, and fragments of the enchrini	The fertile marl lands of Somersetshire. Twerton, Newton, Preston, Clutton, Stanton Prior, Timsbury, Paulton, Marksbury, Farmborough, Corston, Hunstreet, Burnet, Keynsham, Whitchurch, Salford, Kelston, Weston, Pucklechurch, Queenchariton, Norton-malreward, Knowle, Chariton, Kilmersdon, Babington.
17. Marl Stone, Indigo and Black Marl	15		Pyrites and ochre	A rich manure.
18. Red-ground	180		No fossil known	Pits of riddle. Beneath this bed no fossil, shells, or animal remains are found: above it no vegetable impressions. The waters of this stratum petrify in the trunks which convey it, so as to fill them, in about fifteen years, with red watricle, which takes a fine polish.—Highlitdleton.
19. Millstone				
20. Pennant Street			Impressions of unknown plants resembling equisetum.	
21. Grays				Fragments of coal and iron nodules.—Hanham, Brislington, Mangotsfield, Downend, Winterbourn, Forest of Dean, Pensford, Puilow, Chelwood, Cumptondando, Hallatrow near Stratford-on-Avon, Stonebench on the Severn, four miles from Gloucester.
22. Cliff				
23. Coal			Impressions of ferns, olive, stellate plants, threnax-parviflora, or dwarf fan-palm of Jamaica	Stourbridge, or fire-clay.

Mr. Richardson (in 1831) gave the following account of these circumstances, in a letter to Professor Sedgwick:—

"Farley Rectory, near Bath, 10th Feb. 1831.

" Sir,—I am requested to present you the particulars of my acquaintance with Mr. William Smith, well known by the appropriate appellation of ' Strata Smith.'

" At the annual meeting of the Bath Agricultural Society in 1799, Mr. Smith was introduced to my residence in Bath, when, on viewing my collection of fossils, he told me the beds to which they exclusively belonged, and pointed out some peculiar to each. This, by attending him in the fields, I soon found to be the fact, and also that they had a general inclination to the south-east, following each other in regular succession.

" With the open liberality peculiar to Mr. Smith, he wished me to communicate this to the Rev. J. Townsend of Pewsey (then in Bath), who was not less surprised at the discovery. But we were soon much more astonished by proofs of his own collecting, that whatever stratum was found in any part of England, the same remains would be found in it and no other. Mr Townsend, who had pursued the subject forty or fifty years, and had travelled over the greater part of civilized Europe, declared it perfectly unknown to all his acquaintance, and he believed to all the rest of the world.

" In consequence of Mr. Smith's desire to make so valuable a discovery universally known, I without reserve gave a card of the English strata to Baron Rosencrantz, Dr. Müller of Christiania, and many others, in the year 1801.

" I am happy to hear the Geological Society proposes to pay a deserved compliment to his merits, to which I most gratefully bear a willing testimony. I am, Sir, &c.,

" B. Richardson."

For the purpose of giving the widest diffusion to the valuable information, of which the " Tabular View" he had composed was but the index, Mr. Richardson introduced " Strata Smith" to Dr. James Anderson, who was then publishing his " Recreations in Agriculture." This eminent

person immediately added his influence and persuasions to
the many other motives demanding an authentic publication
of views so novel, and bearing so many practical applications, and not only offered the pages of his work for the reception of these views, but proposed a money payment pro-
portioned to their extent. In letters bearing date July 31
and Sept. 12, 1799, he claims the performance of Mr. Smith's
promise, but in vain. It appears from the draft of a letter
in reply to the last, dated Sept. 29, 1799, that Mr. Smith
had been deterred from literary composition by too keen a
sense of his inadequacy to that kind of labour, and had
been disappointed in not receiving some distinct advice on
that head from the learned doctor.

He could " with ease trace each stratum of this country,
from the chalk hills down to the coal," but he would have
had not " the smallest wish whatever to appear in print, if
it were not from a hope that some of the observations might
be of service to the public."

Nothing came of this well-meant proposal, and Mr. Smith
turned all his energies to the prosecution of his profession,
and the tracing out the courses of strata through districts as
remote from Bath as his means permitted him to reach. The
time was favourable. An extraordinary degree of wetness in
the year 1799 had produced, in the vicinity of Bath, an extraordinary phænomenon. Vast mounds of earth displaced
by the augmented force of the springs and the direction of
water into new channels below the surface, were sliding
down the sides of the hills, and bearing to new situations
houses, trees, lawns, and fields.

To remedy such disasters and prevent their recurrence
was exactly what Smith had learned from geology, and had
reduced to practice on many occasions while cutting the
canal. Naturally, therefore, and as a matter of course,
operations of this kind were placed under his care in the
vicinity of Bath and Batheaston; and his reputation for
success in draining on new principles rose daily, and carried
him into Gloucestershire, the Isle of Purbeck, Wiltshire, &c.
Elkington, a Warwickshire farmer, had the merit of ori-

James Anderson, M. D.

from a Sketch by William Smith. 1799.

ginating a system of draining applicable to a considerable class of boggy and springy grounds, and for this he had received a parliamentary grant of 1000*l.* to induce him to discover his *secret.* This system had been tried, and often with entire success ; when it failed, as happened at Prisley, —the bog selected for trial of the process by the Board of Agriculture,—it was because the principles were derived from limited experience, not founded on any general law of the earth's structure. Mr. Smith had found in his geology a truer general theory of springs, a broader and more manageable principle of draining, one which all the experience of his life confirmed and exemplified, and he easily accomplished the drainage of Prisley Bog. Well, then, might he hope for eminent success in this branch of his profession; and viewing it as merely one of many valuable applications in agriculture and commerce to be derived from the establishment of geology, it is not to be wondered at if at this elastic period of life, he thought that, by following out these discoveries through a toilsome and thriftless manhood, he should establish an honest claim for some public provision when his work was done. As a specimen of the fearful state in which the landslips, at this time very common, left particular houses, the following passages are extracted from Mr. Smith's Report to the owner of Combegrove, near Bath (Feb. 1800) :—

" Having minutely surveyed the slips in the ground, and the cracked and dangerous state of the buildings at Combegrove, and made a rough plan of the premises, showing the situation of the respective buildings and the faces of the different rocks, &c. which are affected by this accident, and marked the apparent boundary thereof, so as to bring the whole into one point of view much better than it can be seen upon the ground, I am sorry to find that it is of much greater extent than upon my first survey I had reason to expect, for there is not a building of any sort that has not more or less felt the effect of its fatal consequences. Therefore I am of opinion, that the whole of this beautiful place, in a few years, must inevitably fall a sacrifice to the irresistible

D

pressure of the rocks moving down upon it, if some effectual means are not speedily taken to prevent their further progress." By tunneling into the hill and intercepting the springs, further damage was entirely stopped.

The great humidity of the seasons was followed by a scarcity of corn, and the landed proprietors were strongly aroused to the necessity of providing for the prosperity of British agriculture and the food of the nation, by an improved drainage of the land. So large a portion of the most productive wheat soil of England rests on clay, that generally wet years have been unfavourable to the crop most valued by the farmer. Mr. Stephens of Camerton, the chairman of the canal company, and Mr. T. Crook of Tytherton, one of the best farmers of the Bath district, set the example of encouraging Mr. Smith in his new occupation; and from this time forward, for several years, he was almost daily occupied in various parts of the country, first in *draining* land, and, as a second improvement, very often in *irrigating* it when drained.

From the commencement to the termination of Mr. Smith's engagement as engineer to the Somerset Coal Canal, the remuneration for his time and talents was uniformly one guinea per diem, with allowance for extraordinary expenses. This scale of repayment was continued for some time after the change of his employments produced almost constant travelling; but the numerous simultaneous demands upon his attention, in the new and laborious life which he had entered on, compelled him to raise these terms, from 1801 forward, to two guineas per diem, besides the expenses of travelling. At a later time these fees were again raised to three guineas per diem. Any other than William Smith, equally moderate in his personal expenses, and employed, like him, professionally *for every day* of many years, would at least have escaped poverty; but Mr. Smith, at no time in his life, abounded in money. The principal cause was the liberal, nay lavish, manner in which he expended his means in endeavours to compass his favourite object of completing the " Map of the Strata of England and Wales." For this end he walked, or rode, or posted, in directions quite out of the way of his business; and having thus emptied his pockets for what he deemed a public object, was forced to make up by night-travelling the time he had lost, so as not to fail in his professional engagements. Those who deem such a course imprudent, can scarcely be entitled to censure the motives on which it was founded; his personal loss was the public gain; his individual strength performed a national work; and the sufferings to which this system ultimately conducted, were borne with more than common fortitude.

At Mr. Crook's hospitable house, in 1800, the improvements effected by Mr. Smith's processes of draining and irrigation, were inspected by that prince of farmers, Thomas William Coke of Norfolk, the late venerable Earl of Leicester. This eminent man immediately invited Mr. Smith to Holkham, employed him in a great variety of works, and recommended him to others; and not only valued his abilities for agricultural improvements, but entered warmly into the merit of his scientific discoveries, and conceived an interest in his welfare, which was manifested near the close of his useful and honoured life.

" My journey," says Mr. Smith, " from Bath to Holkham, was performed on horseback, by the guidance of Cary's one-sheet England. The same map was used in my return across the Fens to Peterborough, where I recognized Corn-brash; and, lest I should not recollect the sites, extent, and intricacies of this and the rocks in the series below it, so well known in Somersetshire, I alighted from my horse from time to time as I passed through Northamptonshire by Banbury Lane, and sketched a section of all the ascents and descents on the road, and marked the stone-quarries, outcrops of the rocks, and other strata thereon, and could not refrain from loading my pockets with identifying fossils."

In 1801, amidst incessant occupation and demands for his presence in distant quarters, and even in Ireland, Mr. Smith attempted in vain to commit to paper his fast-growing gene-ralizations in what he regarded as a new science. To this he was stimulated by that never-tiring and always judicious friend, the Rev. Benjamin Richardson, who, in May 1801, alarmed him with the possibility of another publishing those views which should only emanate from himself, and urged the immediate issue of a prospectus and proposals in his own name. This document, which was in consequence drawn up and printed in haste, is now extremely scarce. The title runs thus—

Prospectus of a Work,
entitled
accurate Delineations
and
Descriptions
of the
Natural Order
of the various
Strata
that are found in different parts of
England and Wales,
with
Practical Observations
thereon.

The author evidently proposed to give in this work such a review of his observations as might show the principal facts ascertained regarding the nature and properties of the strata, and demonstrate their practical applications. For this purpose he prepared " A correct map of the strata, describing the general course and width of each stratum on the surface, accompanied by a general section, showing their proportion, dip and direction. The maps and sections, to make them more striking and just representations of nature, will be all given in the proper colours." The following paragraph concludes his prospectus :—

" To attempt a complete history of all the minutiæ of strata would be an endless labour ; for a long life devoted to such a pursuit must be inadequate to the purpose, considering the immense variety which is found in this little island. But should the present essay meet with that liberal patronage from the public which the author has reason to expect, it is his intention, in a future work, to give a particular description of the numerous animal remains and vegetable impressions found in each stratum, with an accurate detail of every characteristic mark that has led him to these discoveries."

" Mitford, near Bath, June 1, 1801."

This prospectus was extensively circulated; and Debrett, opposite Burlington-house, Piccadilly, being named as the publisher, a small MS. map of England, *uncoloured*, was placed in his hands for the engraver. The following letter from his most valued friend, who was in truth his best patron, will explain the then state of progress toward a publication, and demonstrate the earnest and active co-operation of the writer :—

" Bath, July 15, 1801.

" DEAR SIR,—As I may not for some time have the pleasure of seeing you again, I congratulate you on the success of your subscription, which fills readily. And I would have you take Debrett's opinion on the propriety of giving an edition of the work in Latin for the benefit of all Europe, to be circulated under the patronage of our foreign envoys, &c. &c. This would give the system its due importance, and prevent any pirated French edition, which the world will be ready enough to catch at.

   *   *   *   *

" When I rode to the Black Dog, Standerwick, I found the Croydon stone used in polishing wood, at the foot of the sand just above the springs, which may furnish another instance of turning your knowledge to daily use.

" I have left my observations on the strata of North Wiltshire with your brother, and shall be happy if any hints of mine, taken from the Devonshire coast down the Bristol Channel, can be of use to you; for nothing can be more grateful to my wishes than to reflect a single ray of light back to the person from whom I received the illumination, and whose friend and servant

" I most sincerely subscribe,

" B. RICHARDSON."

How much interest the announcement of this work excited may be exemplified by the following letter:—

"Coole, near Gort, Ireland, September 14, 1801.

"Mr. SMITH,

"Sir,—I have distributed your Prospectus amongst my friends, and have the pleasure to request you will add to the list of your subscribers the name of my father, Robert Gregory, Esq., 56, Berners-street, London, and the Hon. Richard Trench, M.P., Spring Garden Terrace, London.

"I find Mr. Evelyn in his 'Terra,' a philosophical discourse on earth, refers to a work of Dr. Lister, called 'A Discourse upon a Map, discovering Sands and Clay, reduced to Tables, presented to the Royal Society.'

"Dr. Hunter in his notes says, 'Dr. Lister was of opinion that, by examining the earth from the surface downwards, as often as an opportunity offered, a pretty just theory might be formed of its contents in general; for it appeared from his own observations, that upper natural soils infallibly produce the same internal minerals and materials. He has thrown out a hint to every naturalist for extending this useful knowledge, by advising that a soil or mineral map should be made, properly distinguished into countries, and enriched with observations for general use, arising from remarks on the bounds and produce of every particular soil. The Doctor thought that sand was once the exterior and and general cover of the surface of the whole earth, and that clay was another coat in the more depressed and hollow parts.' Then follows the table of sands and clay. I hope the above may be of use, and I trust you will excuse me for troubling you with this long quotation from *Hunter's Evelyn's Sylva*, and believe me to be, Sir,

"Your obliged and obedient humble servant,

"RICHARD GREGORY."

"I shall be happy to see you in Berners-street whenever you come to London, or should anything induce you to come over to Ireland (as you mentioned to me at Woburn Abbey that you had some intention to do so), I shall be happy to show you everything that may be the object of your inquiry in this neighbourhood."

In the summer of 1801, Sir Joseph Banks favoured Mr. Smith with an interview, and from this time till his death remained a steady friend and liberal patron of his labours.

In the autumn of 1801, Mr. Smith was, by the kindness of Mr. Coke, introduced to Francis Duke of Bedford, and received from this enlightened friend of agriculture not only considerable employment in draining and irrigation, but special encouragement and assistance in prosecuting his geological researches. After hearing an explanation of the nature and object of those studies, he particularly commissioned his land-steward, Mr. John Farey, to accompany Mr. Smith on an exploration of the margin of the Chalk-hills, south of Woburn, for the purpose of determining there the true succession of the strata, and judging of their most suitable agricultural appropriations, and suggesting means of improvement. Mr. Bevan, of Leighton Beau-Desert, joined himself to this expedition, which was, by direction of the Duke, made at his expense. The following incident occurred in this tour, which took place in the end of January 1802:—

" In this geological excursion, as we advanced near to the foot of the Dunstable Chalk-hills, I ventured on a prediction which in former times might have stamped me for a wizard. I said, ' If there be any broken ground about the foot of these hills, we may find sharks' teeth;' when Farey, presently pointing to the white bank of a new fence-ditch, we left our horses, and soon found six exactly the same as I had seen in 1799, collected by a curiosity-man and antiquary of the name of Yockney, from the chalk-pits above Warminster."

Some further particulars of this journey appear in the following letter to Mr. Richardson:—

" Woburn, 1st Feb. 1802.

" DEAR SIR,—The frost and snow detained me much longer in Staffordshire than I expected, therefore I shall not reach home before the 10th or 12th.

" I have fixed to meet my brother at Down Ampney next Saturday, Sunday, and Monday; thence by Lord Peterborough's, T. Crook's, &c., to Bath, to Longleat and Yeovil, and perhaps down into Devonshire, but must be back into Staffordshire by the end of the month. Now, you know what I have got to do I will tell you what I have done. I have collected a great deal from the North of England and Scotland. Our Mendip limestone, with St. Cuthbert's Beads, goes out to sea at Holy Island, where they are found in great plenty, and are called by this name from the saint of the island *.

" I have found fossils in red marl of Staffordshire, connected some limestones †, and nearly connected some ranges of the coals.

" The last week was spent in surveying distant properties belonging to the Duke, and investigating the country for many miles in different directions as far as Aylesbury, accompanied by the steward (Mr. Farey) and Mr. Bevan of Leighton, I understood by the Duke's desire, to examine the practicability of my arrangement of strata; and though it is a part in which I am the most diffident, they both returned complete converts. Our chief object was the search for coral or other limestones which lie between the sand and chalk hills, which are here situated much the same as in North Wilts. We could not see the oolite, &c., though we saw many marks of it in alluvial gravel.

" But the green sand and Crockerton clay, and several others, were found in many places, the car stone and white sand, &c. &c. &c. But the Swindon stone, with all its fossils, occupies the surface for many miles round Aylesbury, Thame, &c. After spending four fine days very pleasantly in these pursuits, we returned highly gratified and heavy-laden with treasure (not of the East, for that is what everybody knows I was never possessed of, but) such as you and every admirer of nature will be proud to share. Having

* St. Cuthbert's beads are columnar joints of Crinoïdea.
† This passage refers, I believe, to a local patch of limestone in red marl at Arbury in Staffordshire.

Francis, Duke of Bedford

from a Sketch by William Smith

never examined the Swindon Quarries, I was rather at a loss for some time, till I began to recollect the forms of the fossils you have given me. I have no doubt but a little more examination of North Wilts will enable me to arrange every one of these thin strata in the proper order. The great quantity of alluvial gravel which is found in this neighbourhood prevents their being seen on this side of the ridge which divides the drainage of the Thame and Ouse; but this gravel furnished us with many very interesting proofs of a chemical change of substances, which I have long suspected to be occasioned by faults or heterogeneous mixtures near the surface. I pointed this out to Mr. Bevan, who, being a good chemist, satisfactorily explained the cause.

" We not only found decomposed flints and other stones in alluvial gravel, but very luckily met with a place where two known strata were brought together by a fault, which had occasioned a complete decomposition of most of the shells and stones which we had been looking for, and totally changed their appearance; but some of the fossils were sufficiently perfect to recognize their family, comparing them with those we had just before collected. I am now perfectly satisfied of what I have before thought, that the faults in some instances have changed the properties as well as position of the strata.

" I shall be very happy to relate the remainder of my interesting excursion, and remain, dear Sir,

" Your much obliged and sincere friend,

" W. SMITH."

" Rev. B. Richardson, Lisbon Terrace, Bath."

On entering Woburn Abbey for the first time, Mr. Smith was struck by the resemblance of the light-coloured stone of which it is built (from the old quarries of Tattenhoe), to the nearly contemporaneous firestone of Reigate, and began to rub his pencil against its surface, to discover if it was similarly gritty.

The Duke was so much interested in the general progress of these geological pursuits, that, having satisfied himself of their truth and value, he authorized the collecting of speci-

mens of rocks and fossils from all parts of the kingdom, for
the purpose of arranging them in a room at Woburn in the
order of the stratification; and proposed to institute chemical examinations of them. But the bright prospects which
seemed now opening were suddenly closed by the Duke's
lamented death. On the very day, 12th March 1802, which
had been appointed for Mr. Smith to have a second meeting on this subject, the most generous, and apparently the
most judicious of Mr. Smith's noble friends, was buried.

The unexpected death of the Duke of Bedford was
mourned as a public loss; on Mr. Smith it fell with the
weight of a private affliction. Admitted by this truly noble
person, " who valued men much more for their worth than
their titles," (MS.), to unrestrained and friendly intercourse,
he had been successful in convincing him of the real value
to that scientific agriculture which his Grace wished to advance, of a thorough and practical knowledge of the strata
on whose properties so much of the quality of the superjacent soil depends. He had been invited to explain fully the
scientific objects which he sought to accomplish, the progress which he had made, the means which he possessed,
and the assistance which he needed, for completing his
gigantic task. He spoke to a congenial spirit, one who,

"Though born in such a high degree,"

was an honest enthusiast like himself, and had already
thrown all the weight of his rank, intelligence and fortune
into the schemes of agricultural improvement, which then
radiated throughout the kingdom, from Woburn and Holkham. He had found powerful patronage at the very moment when it was most needed, at the time when the public
importance of his past labours was becoming manifest to
the world, and the fearful magnitude of the problem to
which he had devoted his energies began to strike even his
resolute heart with dismay. From this dread the Duke's
kindness had relieved him. " I considered," says he in a
MS. of this period, " my last interview with his Grace as
one of the most auspicious periods of my life. The plan

which he had formed for making himself and others acquainted with the nature of my pursuits was just such as I wished to carry into effect. I had more to expect from his Grace than from all other men in the kingdom."

The effect of the sudden reverse of all these hopes was to delay, indefinitely, the publication of the work announced by Debrett, until the misfortunes of that enterprising bookseller put a stop to the project altogether.

The interest felt at Woburn in favour of Mr. Smith did not, however, expire with Duke Francis. The project of fitting up a geological collection at Woburn, under Mr. Smith's direction, was not altogether abandoned by his brother, who continued to Mr. Smith a high degree of favour and assistance, and entered earnestly into plans for bringing his labours advantageously before the public.

It was high time for Mr. Smith to bestir himself in this respect. In 1802 he was living in Trim-street, Bath, or at least rented a house for the reception of his fossils, and received therein visits from Mr. Wm. Reynolds of Coalbrook Dale, and other eminent persons. Amongst other recollections of this interview with Mr. Reynolds (August 8), Mr. Smith was accustomed to relate that Mr. Reynolds produced from his pocket a copy of the ' Table of Strata' which had been drawn up in MS. in 1799, and already referred to; and stated that, within his own knowledge, copies of it had been sent to the East and West Indies.

There was now exhibited remarkable forbearance on the part of the many eminent persons whose attention was awakened to the examination of the geology of Britain. At this moment, any map, however crude and incorrect, professing to be a mineralogical map of a part of the British Islands, would have been a source of lasting reputation to its editor; any account of the principal facts then ascertained near Bath would have been welcomed with admiration. Had Mr. Smith been exposed to this ungenerous rivalry, he must have sunk under the grief and vexation of being anticipated in his map by some inferior compilation, and in his other labours by notices which, in consequence of his wandering habits and laborious profession, it would have been more easy for others than himself to draw up. But nothing of this kind happened; on the contrary, Mr. Townsend generously presented to his friend a large series of his own characteristic drawings, from fossils in his own collection and other cabinets, as a contribution to his expected work. Mr. Richardson and Mr. Cunnington sent him notes of observations on their journeys; and Mr. Farey, who in this year quitted Woburn, never lost any occasion to advocate the importance and priority of Wm. Smith's discoveries, and to urge their speedy publication. After Debrett's engagement had failed, a new hope sprung up in Mr. Smith's mind, when, besides the Duke of Bedford and Mr. Coke, Sir Joseph Banks and Mr. Crawshay added their powerful patronage of his labours. It was in the latter part of 1803 that professional business at Merthyr brought Mr. Smith and his pursuits to the knowledge of "The Iron King." Accustomed to large views of applications of science, and to a course of action which might be described—

Fortiter in modo, fortiter in re,

this extraordinary man had no sooner made himself master of the peculiarities of Mr. Smith's researches, than he at once resolved to advance the publication of them, not only by assistance in money, and engaging the good offices of his friends, but by inducing an eminent individual to undertake what Mr. Smith's perpetual occupation prevented him from doing, viz. to arrange and prepare for the press his too desultory but valuable papers, which were necessary for explaining the "coloured map of the strata," now frequently exhibited at the numerous agricultural meetings which Mr. Smith attended.

In December 1803, Mr. Smith writes thus to Mr. Crawshay:—"His Grace the Duke of Bedford, and many of the members of the Bath Agricultural Meeting, did me the honour of examining my arrangement of the fossils in the strata. Several agreed with you that it ought to be brought forward as a public thing." These marks of in-

terest were not transient. The Duke's recommendation, in
1804, brought Mr. Smith into acquaintance with Mr. Arthur
Young, then secretary to the Board of Agriculture, and on
the 22nd of May of this year, he had the honour of explaining to the Board the progress he had made in the
maps of the strata, and in the principal applications of geological science to agriculture. He was in consequence requested to draw up some specific proposal for bringing his
discoveries before the public, and to submit the same to another meeting of that distinguished body.

In July 1804*, at the Woburn sheepshearing, Sir Joseph
Banks was present, and listened patiently to a full explanation of the maps which Mr. Smith was exhibiting, and of
the various views of geology applied to practice, which was
the favourite theme of their author. He was at once convinced that the subject was of " such importance, that the
public must have it," and that Mr. Smith's unaided means
would fail to produce either map or book. He therefore
drew up a paper expressing these convictions, to be circulated among the members present, and presented to Mr.
Smith a cheque for 50*l.*, as a moiety of his subscription.
Several other liberal contributions were paid or promised;
but, strange as it may appear, the plan so auspiciously be
gun, was prosecuted no further, though Mr. Smith treasured
the memory of the munificent proposer, and gratefully dedicated to him, ten years later, the large " Map of the Strata
of England and Wales," finished in 1814.

This neglect may be thought censurable; it is, perhaps,
sufficiently explained by the unceasing professional exertions and constant travelling, which left Mr. Smith little
time for more than passing notes of his geological observations, and rendered it an almost impossible task to arrange
detached thoughts into a regular publication. But there
are other extenuating circumstances. His papers were in
London, his fossils at Bath, and he was seldom at leisure to

* In July of this year Mr. Smith was preparing a map of the strata
specially for Sir J. Banks's inspection, and this was probably one of the
documents shown at Woburn.

examine either. While absent from London in the summer
of 1804, a fire happened in Craven-street where his rooms
were, and his papers and maps were removed in hurry and
disorder. To inconveniences of this nature an end was put
by Mr. Smith taking possession of a large house in Buckingham-street, Strand, where his collections and maps were
openly exhibited, and where he for some time employed an
artist in making drawings from his collection for the engraver. In 1805 the Rev. J. Townsend offered his valuable
assistance to throw into form the mass of materials now collected by Mr. Smith (adding also his own excellent drawings), so as to facilitate the publication of them. Of this
Mr. Smith informs Mr. Crawshay in the following letter:—

"June 26, 1805.

"Dear Sir,—I am happy to inform you that my fossils are
at last safely arrived in London, and completely unpacked
and arranged in the same order as they lay in the earth.

"If you or any of your friends are likely to be in town
soon, I shall be happy to be apprised of it, so that I may
have the satisfaction of explaining the subject to you or
them more fully than could possibly be done upon paper.

"I had the honour of showing my maps to the Duke of
Clarence, at the Woburn sheepshearing, and happened to
have my large map of Somersetshire with me, which I have
lately completed, as a specimen of what may be done upon
all the county maps in the kingdom. Sir John Sinclair
thought it a subject of great national importance, and that
the Board of Agriculture should be in possession of those
maps, and that I should be attached to the corps of engineers (who are surveying the island), for the purpose of
connecting my survey of the strata with their large maps.
Sir John wished me to attend the next meeting of the Board
of Agriculture, and says he will do what he can for me. I
should be very glad if something of that sort could be done,
which would enable me to pursue the subject with some
better prospect of profiting by it than I have at present.
Everybody agrees that the mass of information which I
have collected is very great, and likely to be of public uti-

lity; but I find more difficulty in bringing it to market than I expected.

" Mr. Townsend, who has very liberally furnished me with the drawings of all the fossils, has informed me that it will not cost less than three thousand pounds to bring out the publication in two quarto volumes, which he says cannot be sold for less than six guineas.

" The expense of such a publication is too great for my circumstances, and the price of the work will probably preclude many from becoming purchasers; and there seems to be a tardiness amongst many of the great personages who were expected to subscribe very liberally, that makes me loth to engage with proper persons to engrave more of the plates unless I could be sure of defraying the expense.

" I am convinced that every sensible man who shall investigate this subject, must know that a person in my situation has surmounted many difficulties in bringing it to its present state of perfection, and the great loss of time and train of expenses incurred in the course of fourteen years of the best part of my life, which has been spent in this pursuit, has so much injured my pecuniary circumstances, as renders it very disheartening to proceed with a work which still presents nothing but a train of expenses.

" And I have no doubt but you will agree with the rest of my friends, that, if I begin to dispose of any part of it to booksellers or engravers, I may fritter it away till the profit of my labour will be reduced to nothing. Some of my friends have advised me to get out the work in parts or numbers, and I am much inclined to adopt that plan. I have written a new Prospectus, which will shortly be printed, in which I should be happy to have the liberty of making a reference to your approbation of the work.

" I am, Sir,

" Your much obliged humble servant,

" WILLIAM SMITH."

" To Richard Crawshay, Esq.,
Merthyr Tydvil, Glamorganshire."

48

Mr. Crawshay's reply* follows:—

"Merthyr Tydvil, Dec. 27, 1805.

" DEAR SIR,—I have just laid my hand on your letter of
26th June, and am really glad to hear your fossils are all safe
and arranged to your liking in London. I have not visited
town this year, nor do I fix a time for doing so.

" I like Sir John Sinclair's idea of uniting you with the
corps of surveying engineers; shall be very glad to hear it
has succeeded. I do not know what the difficulties are that
prevents your project being brought before the public, except the 3000l. for two quarto volumes, which, if well performed, are invaluable. Supposing the subject explicitly
brought forward, as I expected it would from all the conversation I had with you, I supposed ere this, by subscription, or by your and Mr. Townsend's united exertions, a
work would have been produced that would have promul-
gated a fund of rational amusement and national aggran-
dizement.

" Dr. T.'s† abilities may be inferior to Mr. Townsend's;
but had you followed my recommendation, I really believe before now we should have had a work of great cele-
brity published, from which you would have had the satisfaction of giving to the nation a new idea of raising an
additional source of revenue from the bowels of the earth.
I could dwell on the subject an hour more if my observations were likely to be attended with any advantage to you.
Send me your new Prospectus. If any reference to me will
serve you, you are welcome to do so, but I am too obscure
a person to have influence.

" I did not, at receipt of your letter, notice your wish
therein, or should have immediately replied; and it is by
chance, from Dr. Hall asking me what you had done (or

* The original of this letter was presented to the Society of Arts in
1815, as a testimonial of the opinion entertained of Mr. Smith's labours
by one so competent to judge, on the occasion of the appropriation of a
reward from that Society for the publication of the first geological map of
England and Wales.
† The name of an eminent naturalist.

Rev.ᵈ Joseph Townsend, A.M.

From a Sketch by William Smith.

left undone), that he had heard no more of you. He was,
I believe, with Lord Bute when I sent you to his Lordship.
I sincerely wish you well, and am your friend,   R. C."

The project of uniting Mr. Smith to the corps of surveying engineers alluded to in the above letters failed, probably because it was brought forward at too early a period*;
and from this time ten years more of the most energetic
portion of Smith's life elapsed before the Map of the Strata
(already ten years old) was permitted to appear; the enterprise of a private tradesman (Mr. Cary) accomplishing that
which had been in vain expected from princely patronage
and the sanction of national boards.

In the mean time, however, the spell which seemed to
bind Mr. Smith's energies, and prevent him from appearing
as an author, was broken, by the publication, in 1806, of a
Treatise on the Construction and Management of Water-
meadows. This work, undertaken at the suggestion of Mr.
Coke (to whom it is dedicated) and the Duke of Bedford,
describes in detail the new and valuable processes of irrigation which the author was executing in all parts of the
kingdom, and specially calls attention to the great success of
the experiment on Prisley Bog. It has been long out of print.

The drainage of Prisley Bog (which Elkington had failed
to accomplish) was Mr. Smith's principal occupation during
his first visit to Woburn in 1801. It was an object of great
interest with Duke Francis to conquer this almost desperate
case, and improve the worthless surface of the bog; and
his successor was gratified when Mr. Smith, after thoroughly
depriving the bog of stagnant water, converted it into valuable meadows, by conducting a running stream over the surface. For this excellent work he received the medal of the
Society of Arts in 1805, and a full description of it is in-
serted in the ' Treatise on Irrigation' already mentioned.

These valuable processes of underground drainage and

* Thirty years later the scheme of constituting a geological branch of the
Great Ordnance Survey was revived with better success. This is now in full
operation under the direction of Sir H. T. De la Beche.

surface irrigation were performed (often on the same land in succession) by Mr. Smith in many parts of the kingdom, during several of the following years. Irrigation and draining were in fact directed by no other professional person to any considerable extent, and Mr. Smith's practice extending from Norfolk to Kent, and from Wales to Yorkshire, kept him in continual excitement, and introduced him to the familiar notice of many noble and distinguished persons, who invariably were not only struck by his ability and practical skill, but captivated by his fearless enthusiasm in recommending and following out the study of the ' Stratification.' His time was thus fully occupied in journeys (amounting to 10,000 miles in a year) which enriched his knowledge of the stratification, and in field labours which exemplified the practical value of this knowledge. The details of these innumerable engagements, and others of greater magnitude, are of no interest for the objects of this memoir, except in cases which remarkably extended Mr. Smith's opportunities of observation, or produced results of much public importance.

One of these engagements commenced in 1801, by the request of the proprietors of marsh lands in East Norfolk, between Yarmouth and Happisburgh, which were not only below the level of the rivers, and difficult to be drained, but liable to be flooded by inundations from the German Ocean, through BREACHES IN THE SAND HILLS, AMOUNTING ALTOGETHER TO ONE MILE IN LENGTH! In the face of this dreadful enemy, it was proposed to drain separately the marshes, by embankments to resist the in-rushing sea, and by mills to reject the upland waters; but Mr. Smith, instead of undertaking such feeble projects for particular districts, directly called attention to the necessity of first *stopping out the sea* from the whole vast region of Marsh land. (Report on Hickling Drainage, 1801.) His earnest persuasions, repeated till Oct. 1804, finally prevailed.

The commissioners of sewers took up the matter, adopted Mr. Smith's plans, and accomplished, almost in a single summer (1805), the expulsion of the sea from seventy-four parishes in Norfolk and sixteen in Suffolk, which, by an act of James I., 1610, entitled " The Norfolk and Suffolk Sea Breach Act*," had been declared liable to contribution.

The district of country to which these remarks apply, resembles almost exactly the portion of Yorkshire called Holderness. It is a complicated area of rather elevated clay lands (locally called ' hard' lands), ramifying or rising in detached mounds within an expanded and tortuous surface of 'soft' marsh lands, below the level of high water, and commanding only a very limited fall to low water. These marshes, naturally rich and improveable, capable of yielding excellent grass land, in a county where natural pasture is valuable, were undrained and unimproved; nor until the sea could be effectually kept out was it to be supposed that any improvements could be attempted. The sea coast offered a natural barrier, which can be easily described. Where, as north of Happisburgh, and for two miles to the south of it, the ' hard' land comes to the sea, its cliffs of diluvial mud are slowly wasted and fall in ruins. This hard land retires from the sea about a mile from Happisburgh lighthouse, and the marsh lands succeed upon a long range of coast, scarcely interrupted by the detached hill of Palling, to Winterton. Along all this coast the sea might enter, and spread in broad and winding sheets over 40,000 acres of land, but for a natural barrier of sand-hills, thrown up into a narrow irregular ridge by the action of the sea and the wind, and fixed by the growth of the ' Marram,' or *Arundo arenaria.* The set of tide along this coast is from the N.W., and this being the line of the mud-cliffs and sand-hills, the whole of the sea-barrier thus described is raked by the currents, and the materials of which it is composed are perpetually drifted to the S.E., to augment the mass of sand-banks about Yarmouth

* The act recites, that " within six months now last past the sea hath broken down and washed away the cliffs and higher grounds there, such as they were, and laid them flat and level with the said inlands, and made breaches so wide, that the sea hath broken in at every tide, and with every sea-wind, into the very heart and body, as well of the said county of Norfolk as into some parts of the said county of Suffolk adjoining, which be subject to the said overflows"—" being grounds of themselves very rich," &c. &c.

Haven. The wasting of the cliffs to the north supplies the materials for the aggregation and renewal of the ' sand-hills,' also called ' meal hills*,' or ' dunes,' or ' denes,' and the whole coast is in motion. So long, however, as the sand-hills maintain a continuous unbroken line, they offer an effectual though variable barrier to the sea; what they have of this continuity and integrity is owing to the growth of the valuable plant already named, for its roots spread amongst and bind together the sand, and its ' bents' (stalks) check the devastating action of the wind. Placed thus in un-stable equilibrium, the state of the sand-hills at any moment expresses the balance of the integral effects of the sea-wind and vegetation, and from time to time this balance is un-favourable to the safety of the inland country; for the wind often conquers the marram, by heaping up sand more abun-dantly where this plant grows the best; inequalities are thus occasioned; tempests succeed; the relatively depressed parts of the sandy chain yield to the wind, receive the spray, and admit unusually elevated waves of the sea. ' Gaps' or ' breaches' (fearful name!) are thus generated; the ocean, swelled by north-west winds, rushes in; the internal rivers are choked by the simultaneous flooding of the Yare with salt water; the marshes are drowned, and years pass before the soil recovers its natural state. The processes thus sketched had gone on with various fluctuations and excited frequent alarms, old breaches being filled by nature and new ones opened, as the conflicting causes predominated. In 1792, Mr. Faden measured nine gaps, amounting to 484 yards, opposite Horsey; but in 1805 there were new ones opened, and the whole, between Winterton and Happis-burgh, measured nearly one mile!

The way in which Mr. Smith set about the repairing of these broken hills was eminently characteristic of his quali-ties as an engineer. After considering a variety of plans which had been proposed for stopping the breaches by timber! by stone! by clay banks! &c., he examined the operation of the tides and storms on the coast, compared

* From the Welsh ' moel,' a round hill?

the levels of the high and low parts, and finally proposed to make all the new artificial embankments as like as possible to the *natural embankments* thrown up by the sea on the same coast, to make them of the same materials, and to give them such directions as might best shelter the new work by the old. A plan so simple was almost rejected with ridicule, till, by walking on the sea-shore and pointing out to his amazed companions how ineffectual and short-lived was the resistance offered by solid constructions to the rage of the sea, and how permanent was the power of sloping banks of sand and pebbles, in particular directions, to exclude the ocean which had formed them, he convinced the most sceptical, and compelled the most contemptuous to exclaim, " Oh, that none of us should have thought of this before!"

The plan was indeed simple enough, and required almost nothing but labour for successful accomplishment. Mr. Smith chose slopes against the sea as moderate as any part of the natural beach on that coast, viz. about twelve yards horizontal for one yard elevation, the back or inland slopes being steeper (four or five to one). By watching the aggregation of sand and pebbles on the shore, he found that, at particular seasons and by unusual storms, the bed of the sea was disturbed, and the sand became covered by pebbles or ' shingle' scattered with much uniformity. These shingle-beds were effective in binding down the sand which would otherwise have drifted with the wind, and he resolved in this respect to imitate his great teacher—Nature. Accordingly carts in great numbers were employed in removing sand and making great mounds across the gaps, and then, especially when the tides threw up ' shingle,' the sandy bank was sealed down with a bed of pebbles. On these unresisting slopes the mightiest storms of the German Ocean now break harmless, and a very slight annual charge is sufficient to maintain the form and substance of the work.

It is needless to enlarge on the sagacity which thus invented the simplest of all possible remedies for one of the most dreadful of all evils. Nature's violence has here been conquered by implicit obedience to her immutable laws; the

means of almost unlimited improvement of a vast breadth of
good land are in the hands of the owners ; a great boon has
been conferred on humanity, and a great example set of the
benefits which may be derived from a faithful observation of
nature.

After the sea breaches were effectually stopped, Mr. Smith
was able to suggest to the proprietors of the marsh lands
which had been thus benefited, effectual methods of draining (chiefly by mills throwing up the water into rivers) and
improving them.

In superintending the sea breach repairs and regulating
the assessments to pay for this work, and in the construction
of water-meadows for Mr. Coke and other persons, Mr.
Smith spent a portion of every year, from 1800 to 1809, in
Norfolk and Suffolk ; but other occupations were continually
requiring his presence in all parts of the kingdom, and gratifying him with new views of the strata.

In Sept. 1802, we find him examining the ancient slaty rocks
of Wales in the vicinity of Dolymelynllwyn near Dolgelle,
on occasion of a visit to inspect Mr. Maddocks's successful
embankment across the Traethbychan at Tanyralt (Trema-
doc). He also ascended Snowdon with guides furnished by
Mr. T. Asheton Smith, examined the copper mine then at
work, and studied the slate quarries of Llanberis. Of this
journey he retained, in after years, the most vivid and pleasing
recollections, describing with enthusiasm the grandeur of
Cader Idris, and the magnificent view from the summit of
Snowdon at the rising of the sun.

In 1803, and for several years following, we find him engaged as a mineral surveyor in Yorkshire, Lancashire, South
Wales, Somersetshire and Gloucestershire, establishing a
colliery at Torbock near Liverpool, directing a trial for coal
at Spofforth, and examining the geology of Witton Fell in
Yorkshire ; investigating the outcrops of coal in the vicinity
of Newent, the Forest of Dean, Nailsea, and Kidwelly ;
and constructing sea-banks at Laugharn. In 1809 he began
to execute the Ouse Navigation in Sussex ; in 1810 he restored the hot springs of Bath which had failed ; in 1811 he

began to examine into the causes of leakage on the Kennet and Avon Canal, and in his report on the best method of obviating this mischief proposed a new set of operations to supply the canal with water; reported on trials for coal in Buckinghamshire; planned and surveyed improvements in Kidwelly Harbour; in 1812 executed the Minsmere drainage in Suffolk. These and a hundred other professional engagements, while they furnished the means and occasion for that incessant travelling, without which no single man could have accomplished so great a work as the ' Map of the Strata of England and Wales,' left him no time to prepare either maps or books for the press.

Careful inspection of the enormous load of maps and papers relating to this period of Mr. Smith's history, has, however, demonstrated that between 1799, when he first contemplated a publication, to 1812, when this object was fairly taken in hand, Mr. Smith was acting upon the Linnæan maxim,

"Nulla dies sine linea,"

for, at almost every point of rest in his undulating and rapid career, he committed to paper and dated some note of the day's observations of the direction, dip, and aspect of the rocks he had passed over. The following is a specimen of such observations at an early period of his inquiries :—

" Feb. 6th, Cirencester to Bath, 1803.

" In crossing the Cotswold Hills from Cirencester to Bath, we are above the upper rock to the thirteenth milestone, and in part of this space, from Tetbury to Didmarton, the stuff above it occupies the surface in many places, and produces wheat and pasture-land, with plenty of water, slate and wood, which give to the vicinity of Tetbury quite a different appearance to many other parts of the Cotswold Hills. The most perfect part of this upper stratum seems to be about Doughton or I. P. Paul's, and a summit-surface then extends towards Beverstone, which seems to be free from the common intersections even of dry valleys, and continues to Kingscote, and so on in one connected ridge of high ground till it terminates in the bold promontory called Stinchcombe Hill. This point, I believe, is composed of under rock and its accompanying sand; but I am inclined to think that the upper rock reaches as far as Kingscote, or perhaps to the Gloucester road, which is some distance beyond it. The clay and fuller's earth which lie between the two rocks evidently produce the pasture-land which is about Nimpsfield; and this place is supplied with water from the springs which these strata generally produce. All the streams which flow from this ridge through the many deep vales in different directions must have their source from the same stratum, and I expect that this will be found to be the case with every stream on the western side of the Cotswold Hills from Minchin Hampton to Hawkesbury-Upton, where the two rocks will be found to come very near together in the same hill which constitutes the high ground from thence towards Horton. The outlines of the two rocks then recede from each other again and become much more distinct than I have yet found them on any other part of the Cotswold Hills. The upper rock between the thirteenth and fourteenth milestone dips very considerably towards Badminton Park, and has a very apparent outcrop in the dry freestone land near the thirteenth milestone; the fuller's earth and a bastard blue clay, and the usual accompaniments of those strata*, then plainly appear by the new plantations going down the hill, and may be traced by the coldness of the surface and outline of the upper rock, which, with two openings through it, runs in a line very near to the Cross Hands and the turnpike road very near Mr. C.'s park. It then returns toward Tormarton (where it appears to have an opening through it), runs out in a ridge at the south side of the park, has an opening through it in the dip, which gives rise to a valley. Cold Ashton, on the opposite side of this vale, plainly points out

* By the 'usual accompaniments' are probably meant the usual fossil shells, especially *Ostrea acuminata*, here plentiful in the strata designated. By 'upper' and 'under' rocks, the two oolites of Bath are designated. The description here given of the ranges of these rocks is remarkably exact, as I learn by Mr. Lonsdale's Map, and by the progress of the Ordnance Geological Survey in this district in 1843. The geologist who shall be at the trouble to read this somewhat dry description of observations in the field will require no further evidence to recognise in the author a 'great original discoverer in English geology.'

the line of its corresponding outcrop, which runs round the point of the hill very near the cross ways on the road from Marshfield to Bristol, then returns by the line of the cold wet lands on the south side of the ridge between Cold Ashton and Marshfield. Insulated parts of the upper rock very evidently appear upon the east end of Huntrick's Hill, Charmy Down and Little Salisbury, and to the left run out in another broader part, which includes Toll Down, makes another longer return toward Littleton, and leaves a narrow opening between this line and its opposite outcrop, which stretches from Littleton through Littleton Wood, and, crossing the road by the turnpike, runs out a little way to the right, and returns again to another vale between the parting of the roads and Cold Ashton. The main line before described from Cold Ashton to Marshfield continues in a connected line, which is deeply indented by the many ramifications of the Catherine Stream until it joins the high land at Culeron Down. The crop of this rock may then be traced round by the high part of Mr. Wiltshire's land at Culeron, &c. &c., until it joins the before-mentioned bank on the north side of Cold Ashton, making one connected line all round it, which renders this piece of the upper rock as completely insulated as that on Lansdown or any other of the smaller hills. If we now go back to the thirteenth milestone and look over the face of the country again. it will be readily perceived that the clayey strata of fuller's earth, &c., before described as the source of streams on the western side of the hills, have now become the source of streams which run in a different direction. The sharp rise of the under rock to the west, and the outline of the upper one to the east, before described, form the first outlines of these little hollows, many of which (though dry) may be properly called the primitive source of brooks, rivulets and rivers."

Whenever opportunity offered, he possessed himself of records of borings, natural and artificial sections, drew them to a constant scale of eight yards to the inch, and coloured them. From a mass of these sections, a few are selected. The following account, drawn up in 1805, of the strata in the now almost forgotten coal-field at Newent, may perhaps

58

be consulted with advantage, if, as there may be reason to
advise, the gentry of that neighbourhood should resolve to
make an experiment to reach at a greater depth the coals
which were anciently worked near the surface.

BOWSDEN COALWORK 1 MILE S.W. OF NEWENT.

	Fathoms.	Yds.	Feet.	In.
1. Gravel	2	1	0	0
2. Marl	1	0	0	0
3. Red and White Rock	3	1	0	0
4. Duns, &c.	8	0	0	0
Coal.				6 ft. in 4 coals.
5. Duns, &c.	4	0	1	6
II. Coal.				3 ft. 4
Strata	19	0	1	6	9 ft. 4 coals.
Coals	1	1	0	4
Total	20	1	1	10

The manner in which Mr. Smith combined his measures of the strata near Bath, may be understood from the following sections, dating about 1800.

SKETCH OF STRATA ABOVE THE GREAT OOLITE.

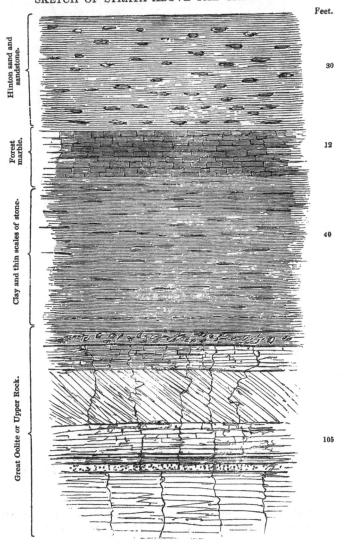

UPPER PART OF THE GREAT OOLITE.

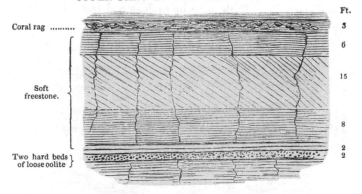

Coral rag ....

Soft freestone.

Two hard beds of loose oolite

Ft.
3
6
15
8
2
2

The oblique stratification represented in these sections is of frequent occurrence near Bath, and in Gloucestershire.

COMBE GROVE PIT.

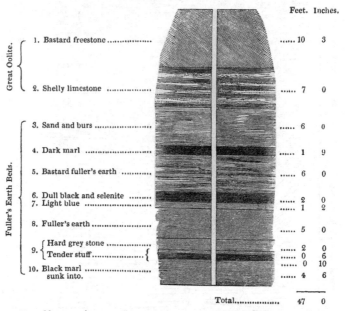

		Feet.	Inches.
Great Oolite.	1. Bastard freestone	10	3
	2. Shelly limestone	7	0
Fuller's Earth Beds.	3. Sand and burs	6	0
	4. Dark marl	1	9
	5. Bastard fuller's earth	6	0
	6. Dull black and selenite	2	0
	7. Light blue	1	2
	8. Fuller's earth	5	0
	9. { Hard grey stone	2	0
	{ Tender stuff.................... {	0	6
		0	10
	10. Black marl	4	6
	sunk into.		
	Total..................	47	0

Profile sections were sketched on most of the roads which Mr. Smith passed over, such as that given on the next page, date 1805.

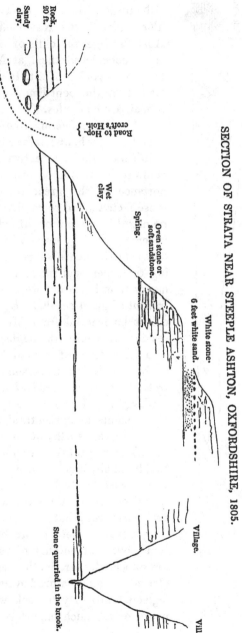

SECTION OF STRATA NEAR STEEPLE ASHTON, OXFORDSHIRE, 1805.

SKETCH OF STRATA ON THE SHORE OF THE MENAI,

Between the Salt Mill and the Limekilns belonging to T. Asheton Smith, Esq., 1802.

Salt Mil.

Road.

Lime-kilns.

Natural cliffs were always copied. The faults in the New Red marls at Aust Passage, the cliffs of 'red rhab,' as he considered them, at Minehead, the 'red rhab' of St. Ishmael and Laugharn, the bent limestones on the Menai, are examples.

While thus collecting from all sources a body of authentic information, Mr. Smith was still more active in endeavours to impress upon others the importance of the great truths which these facts tended to confirm. In this he might be often thought tedious, but never presuming, even when he addressed to his hearers arguments touching their personal interests; and remarks thus hazarded were sometimes fruitful of great results. By mere accident he learned from Mr. P., then resident at Tenby, the neglected state of a large tract of ground belonging to that gentleman in the south-west of Ireland, *on which there had never been a plough*; and after in vain suggesting attempts at agricultural improvements, as well worthy of his attention, and likely to gratify as well as benefit him, he finally inquired if no subterranean treasures had ever been suspected on the estate. He was carelessly answered, that at some unknown time previous, metallic ore had been found, but that nothing of value was now expected. Mr. Smith paused, and advised the good-natured owner of this neglected property to look well to a matter of so much importance. The consequence was, the establishment of the now famous and very profitable Alighies Copper Mine.

While engaged in the superintendence of the Ouse navigation (1809), Mr. Smith was very frequently at Tilgate and Cuckfield, and obtained from quarries at this last place several bones of gigantic dimensions, which, with the rest of his collection, were transferred in 1815 to the British Museum.

The following notice on the geology of Sussex is worth preserving :—

"Scarborough, June 30, 1839.

" I became engineer to the Ouse navigation (the works on which had been abandoned about fourteen years before), and under my management it was extended from the vicinity of Sheffield Place up to the Balcombe Road. Here I found in the names of ' Hammer-ditch,' ' Furnace-pond,' Cinderhills,' &c., the traces of iron-works anciently carried on to a great extent in Sussex : indeed, an elderly gentleman who was one of the commissioners on this business had been one of the Sussex ironmasters.

" In searching for stone to build the locks and bridges, and by various geological excursions, I became acquainted with the strata, and collected many of the Sussex fossils, some of which were left unnoticed in my Stratigraphical System, from scarcely knowing what stratum they belonged to, and which I think is still dubious in much of the interior of the Wealds.

" I was, however, sufficiently well acquainted with the stratification to draw its great outlines on the large county map for my friend Mr. Isaac Watt. This, prior to the Ordnance Survey, was considered to be one of the best county maps in England, patronized by, or made at the expense of, the Duke of Richmond.

" There being about this time a projected canal through the northern side of the Wealden district surveyed by Mr. Rennie, I took the levels for a line to connect this with the Ouse navigation up the Balcombe Valley, by a tunnel through the forest ridge, and spent some time unprofitably in preparing a plan of it, which was deposited with the Clerk of the Peace at Lewes."

In 1810 the Bath hot-springs had failed, and Mr. Smith was hastily sent for to restore the water to the Baths and the Pump-room. This alarming circumstance, already nearly forgotten at Bath, put in exercise Mr. Smith's peculiar abilities and patience. Not without much opposition, he was allowed to open the hot-bath spring to its bottom, and thus to detect the lateral escape for the water. The spring had in no sense failed, but its waters flowed away in new channels. The men, in the excavation thus opened, found the heat oppressive; wax candles were employed for illumination, and the gushing water raised the thermometer of Fahrenheit to 119°. In the channel of the spring was found a bone of some ruminant coated with crystallized sulphuret of iron, and a rolled flint of the kind commonly seen in the Wiltshire chalk hills, and, like them, full of spongoid organization. The operation was perfectly successful, and the cure complete, the baths filling in less time than formerly.

The failure of these springs was by many persons attributed to the sinking of a pit at Batheaston, which at this time was in progress. The pit was sunk as a trial for coal, after several years' discussion; and it happened that, at the same time that the Bath Corporation required Mr. Smith to restore their springs, he was consulted by the projectors of the 'Batheaston Colliery' as to the best means of overcoming a mighty influx of water into the pit. It was said, probably with truth, that the water which entered Batheaston pit at the bottom was of elevated temperature; and many persons connecting in imagination these opposite events, which occurred at the same time, loudly expressed their conviction that the Batheaston pit robbed the wells at Bath, and ought to be filled up. A coal-pit at Batheaston would have been a poor equivalent for the loss of the great attraction to Bath; and though the fears of a total failure of the hot water might be groundless, still no one could pronounce that the springs of such a nature, and rising evidently from a great depth, *perhaps on the line of a fault not*

discoverable at the surface, might not be affected by a pit
sunk three miles off. Perhaps these views really influenced
the fate of that once favoured coal experiment. Mr. Smith
found the means of plugging the bore-hole at the bottom of
the pit through ninety yards of water, and actually drained
it; but the undertaking was abandoned, and only a feeble
weeping of water now marks the place where this interesting
enterprise was begun.

The following remarks were written in 1832.

"Hackness, Nov. 8, 1832.

"But now to reminiscences in the vale of Bath. The
Bath Avon in its way to Bristol crosses two distinct and considerably distant exposures of the coal-measures, and two
of the lias and two of the red ground, so that Tiverton
Mills and Keynsham Bridge are both upon lias; therefore,
admitting the river to be an inclined plane with about forty
feet fall, the stratification in the bottom of the valley is evidently in a greatly distorted state.

"Portions of each of these strata are considerably higher
on each side. The lias at Weston is higher, lower toward
the Crescent fields, and by declination south and eastward,
Bath Bridge is said to stand upon beds of that rock.

"The hot-springs in the city rise to the surface through
blue marl or lias-clay, their issue not more than fifteen or
twenty feet above the river. The discharge of hot water
at all the springs is said to be five hogsheads per minute.

"Batheaston Pit, ninety yards deep in this clay and stone,
may be forty or sixty feet higher.

"The water from the lias rock in the pit ascended to the
surface, and, with what was let up by boring the red rock
beneath, stopped a steam-engine of eighty or ninety horse
power.

"From these known circumstances at Batheaston, we may
infer that the Bath hot-springs have a deep source, and
rise as through a natural bore-hole in the blue marl to the
surface. Both these great discharges are on the same side
of the Avon, and both, of course, receive their supplies of
water from strata which, at a high angle of elevation, rise to

F

a considerable altitude in the hills north thereof. The coal measures do not appear in these hills ; but beyond Lansdown Hill, and only four miles north of Bath, coal has been wrought at Wick, near the mountain limestone, which lime-stone appears in scars still higher east thereof. In the deep bottom, by Hamswell House, west or north-west of Bath-easton pit, but much higher, there is an outburst of the conglomerate portion of the red marl rock called millstone, so that conductors for water in abundance towards Bath and Batheaston are not wanting; and hence, if the coal-measures should thereabout be unconformably covered, it is possible, by a singular combination of circumstances, that the Bath waters may be a compound from the lias, red rock, coal-measures and mountain limestone."

The diffusion of correct geological principles has nearly removed from the chronicle of passing events the records of trials for coal in districts of Great Britain where nature, in-terpreted by science, forbids the discovery of that inestima-ble treasure. In the early part of Mr. Smith's career such experiments were common ; on one occasion he paid twenty-four shillings for a parcel of micaceous sandstone, taken from the old red formations of Herefordshire, and put in *the mail-bag*, as specimens of the matter sunk through in a trial for coal! In the vicinity of Wincanton, at a time when he was professionally employed in that country, there was in pro-gress a ridiculous project of this kind at Bruham ; and, in spite of remonstrances from Mr. Smith and his intelligent friends, the speculators proceeded at a ruinous expense through the clunch (now called Oxford) clay, Kelloway's Rock, with "the small lobate oyster" (*gryphæa dilatata*), till they entered rocks of the oolitic series. At Bagley Wood, near Oxford, within sight of the university halls, which then resounded with the fame of the attractive and useful lectures of Buckland, an absurd experiment for coal was begun in the Kimmeridge clay, and ended in a deplo-rable sacrifice of fortune.

In 1811 Mr. Smith was called in to inspect and report upon a singular series of such trials on the estate of Sir

John Aubrey, at Brill, and had the melancholy satisfaction of proving, by general geological truths and by special examination of the facts disclosed by the boring-rod, that the trials ought never to have been commenced. Strange that, while so many persons of easy faith should be persuaded by idle conjectures and deceitful analogies to follow the guidance of " a practical man," and search for coal where it is not, others are so firmly incredulous that they deride all the positive facts and reasonable arguments advanced by " men of science," which prove the probability of obtaining coal in a variety of situations yet untried, and determine the best localities for the experiment!

There is another subject of importance, which baffles not only the " practical man," whose rules of conduct are merely local, and only useful in cases strictly parallel and known to be so, but even the scientific engineer of more enlarged views,—the course of subterranean drainage, and its effect in producing land-slips, leaks, blow-holes, and breaches in canals. Canals which cross the oolitic ranges of hills are particularly liable to these evils, from the regular alternation of thick open-jointed rocks with impervious clays, and the complicated sinuosities and denudations of the strata; for the consequence of these circumstances is the pouring out of innumerable springs in ranges along the hill-sides, where the strata are regularly inclined, and in confused groups where they are " faulted" or " slipped." The action of the springs themselves contributes to generate such slips, and to continue for a long time the unquiet state of the ground when once it has been put in motion, so that insecurity is almost perpetuated. Moreover, the canals which cross such a country are cut alternately through open-jointed and even cavernous rocks, and closely water-tight clays. In the latter strata the excavation itself is usually almost water-tight, but in the long passage (which sometimes amounts to several miles in length) across the limestone rocks, all the skill of the engineer, supported by even lavish expenditure, often fails to prevent the water from running away as fast as it enters. Such canals, if they happen to have their " summit level" (as the

Thames and Severn Canal at Salperton Tunnel), in these rocks, are like the buckets of the Danaids, and with the water goes the profit. In vain the Thames, raised from its source by a mighty engine, is poured into such a thirsty canal, the flood passes into the gaping rocks below, in spite of renewed puddling and continual repairs. But this country presents still another difficulty, in consequence of the very frequent cases of slight derangements, small faults and slides, of portions of rock, clay and sand, at the edges of the hills. These displacements confuse the natural direction of the springs, and enable them to destroy the resisting power of the artificially " puddled " canal. The springs thus circumstanced, often short in their course and temporary in their flow, fill or leave empty the subterranean chasms in the ground that has moved, wash away sandy parts, press upwards against the canal base, or laterally against its sides, weaken insensibly its defence of artificially compacted clay, cause a " leak " or a " breach," and let through with violence a portion or even all the body of the water contained in that level of the canal.

Inconveniences and losses from these causes on the Somerset Coal Canal, which Mr. Smith planned and executed across the oolitic ranges, were in a great degree prevented by the application of the general geological views which he possessed, and which more than supplied the want of that experience which no one possessed ; but the same causes produced disastrous effects on the Kennet and Avon Canal, to which the Coal Canal is joined between Bradford and Bath. In 1811, the evils thus occasioned had grown to be of serious magnitude, and Mr. Smith was ordered to report on the subject of the springs in this portion of the canal with a view to an efficient remedy. The Rev. Mr. Townsend, a member of the committee of management, took great interest in the investigation, and communicated to his old friend a remarkably clear summary of the evils, and some valuable suggestions for overcoming them.

In his Reports on this subject (1812 to 1814) Mr. Smith traces the natural circumstances above referred to, which, by their unfortunate combination, foiled the skill of the ablest engineer of his day (Rennie), and rendered the masterly works which he planned subject to grave accidents, from which, in any other country, their admirable execution would have entirely preserved them. By help of a geological map of the line of the canal, he points out the circumstances which rendered the " leaks and blows fatally common in certain parts of the canal," and notices the minor but not-to-be-neglected influence of plants (like *Equisetum palustre*) and small burrowing animals, which, by penetrating the banks or bed, often originate leaks. Where the canal crosses the loamy sand below the inferior oolite, the canal is in danger; for, in the first place, that sand is very apt to become cavernous by the passing of water through it (as through the " fox earth" of Dorsetshire), and thereby endanger the puddling; and next, when employed in the banks (or left without puddling on the bed), it is apt to be the favourite habitat of *Equisetum palustre*, whose deep roots penetrate and render it unsolid, and the resort of small animals, which perforate to the water. Where the canal crosses the jointed rocks, these give passage to water at one time and to air at another, whenever they are the channels of merely temporary streams, fed by rain, and ceasing with drought. What can be more destructive to the puddles of a canal than to have them washed in the winter by currents of water and dried in the summer by currents of air? If the canal from any cause becomes short of water, its banks, on their internal surface, are liable to crack, and thus the evil is augmented; and though great streams should be poured into a canal subject to leaks on its bed or sides, this excessive supply may even augment the evil by increasing the waste and friction through such leaks.

The general remedy for all this is the entire interception of all the springs which rise from a level above the canal and pass below it through natural fissures and cavities. This is a process requiring great skill and extensive experience; some of the springs which it is most important to intercept come not to the surface at all in the ground above the canal, but flowing naturally below the surface through shaken or faulty ground, or along masses of displaced rock which extend in long ribs from the brows down into the vale, emerge or attempt to emerge in the banks of the canal; these no ordinary surface-draining will reach, and none but a draining engineer, well-versed in the knowledge of strata, can successfully cope with such mysterious enemies. But Mr. Smith, confident in his great experience, not only proposed, by a general system of subterranean excavation, to intercept all these springs, and destroy their power to injure the canal, but further, to regulate and equalize their discharge, so as to render them a positive benefit. This he would have accomplished by penning up the water in particular natural areas or pounds which really exist in that and other countries between lines of " fault," or certain ridges of clay (" horses") which interrupt the continuity of the rock and divide the subterranean water-fields into limited districts, separately manageable for the advantage of man by the skilful adaptation of science. Such a noble project was then and is now little adapted to meet the views of ordinary and uninstructed men, but it has been exemplified on a small scale by arrangements designed by Mr. Smith for supplying Kirkby Lonsdale and Scarborough with water, and is really a simple and easy deduction from geological data *. Its importance will become evident at a future time; but to embody a valuable theoretical truth in wise and economical application belongs only to minds of a high order, and for these (in the present state of public opinion regarding the value of science) the quiet fields of philosophical research are more attractive than the struggle for practical results, against opposition and prejudice, in the toilsome arena of daily life.

We are now arrived at a critical period of Mr. Smith's career, the recommencement of his efforts for producing a part of the great work on which he had for twenty years been occupied. " The end of 1812 brought me a proposal from Mr. Cary to publish my Map of the Strata. Terms

* See an account of the arrangements at Scarborough, drawn up by Mr. Smith, in Phil. Mag. 1827.

were soon settled, and the work commenced with the beginning of January 1813." (MS.)

On looking back through several years of his professional toil, it might almost appear as if, while devoting the profits of that toil, and selling his small patrimony to eke out his earnings, in order that he might be enabled to visit and examine the most remote parts of the island, Mr. Smith was forgetful of his own reputation and his promises to the public, and intent only on gratifying at all risks an indomitable and feverish desire to observe and know the structure of his country. This supposition would, however, embrace only half the truth: it is true that Mr. Smith was *discouraged* by several events, such as the ill success of the far-advanced scheme of publication with Debrett; the loss of the Duke of Bedford; the unsuccessful, or at least inconclusive appeal to the Board of Agriculture; and the death of Mr. Crawshay. But yet he was constantly endeavouring to burst the gloom which straitened means and wearisome business spread over his prospects: he employed an artist to draw, and another to engrave select examples of his fossils, and coloured several copies of maps and sections in hopes of some favourable occasion to bring them forth. Meantime circumstances of great importance and significance had happened. In 1807, the Geological Society of London was formed, and several of his earliest friends were admitted honorary members thereof, while he was almost unnoticed, except by visits from Mr. Greenough, the President, Sir James Hall, and a few other members, to examine his collections in Buckingham-street (March, 1808). Mr. Sowerby began (1808) to prepare drawings for his work on the Mineral Conchology of Great Britain; Mr. Farey, having gained the patronage of the Board of Agriculture, proposed and executed a Mineralogical Survey of Derbyshire (1811); and in 1812 appeared Mr. Townsend's ' Character of Moses.'

Notwithstanding these exciting motives to activity, Mr. Smith's own pecuniary means were really so feeble, even aided as they had been by the constant liberality and encouragement of Sir Joseph Banks and a few other friends of science, that the preparation of a great and costly Map of England and Wales was entirely beyond his power, and his spirit must have sunk in unequal conflict with accumulating difficulties but for the courage and resolution of Mr. Cary. The maps of this eminent engraver had for a long time enjoyed the highest reputation, and it was upon the large map of England which he had published that Mr. Smith had collected and arranged the fruits of twenty years' inquiry, *marking only the exact points, or drawing only the limited lines which were really ascertained.* On these maps, full of names and designations of political geography, colouring of the lines of strata was impracticable; and Mr. Cary undertook the drawing and engraving of a new and very elegant large map*, which, instead of political divisions, should be richly marked by names of natural districts, and by a full delineation of the innumerable small streams, whose distribution is an important feature of physical geography. Both in respect of these names and the terms to be employed in distinguishing the groups of strata coloured in his map, Mr. Smith was a good deal embarrassed, as the following notice shows:—

"Scarborough, May 17, 1839.

"For several years after the foundation of the earth's history was securely laid, we had no words for the science, no language in which we could convey our ideas; its present comprehensive name of Geology remained unnoticed in dictionaries and unuttered in England, and usage had scarcely settled whether the word strata should not have an *s* appended; but how numerous are now the words from the dead languages which geology has revived and brought into common use all over the world!

"Much doubt remained for a long time whether the science, like chemistry, should not have a language of its own; and I, so very incompetent to the task, thought much about a new nomenclature, and have been at different times strongly urged to it by deep-learned men; but having

* In fifteen large sheets, altogether 8 ft. 6 in. high and 6 ft. 2 in. wide.

dictated, off-hand, in the plain language of the country, a tabular view of the science to my two first pupils, the Rev. Benjamin Richardson and the Rev. Joseph Townsend, that crude manuscript, without any revision whatever, was faithfully transcribed from one to another, and soon despatched to remote parts of the world.

"The new cultivators of the science found, as I had done, the necessity of accommodating their language to those in the country from whom they had to collect the facts; and so, in transmitting by the press the knowledge acquired, some old Saxon and British words have been brought into use; the further advantage of the science in explaining the many descriptive names of places, and the *circumstantial history of Britain*, is yet to be developed. I had long been doing much in this way, when circumstances (about 1813) gave me the advantage of great aid from my learned and good friend Henry Jermyn, Esq., of Sibton Abbey, who was then compiling a history of Suffolk.

" We fortunately became coadjutors in draining the great level of Minsmere Marshes, he being one of the Commissioners and I the engineer.

" The writing on my great map being then required by the engraver, I told him my views of introducing, as far as possible, only the most descriptive names, which, to him, was a new view of their derivation, and we presently went to work in his library; I having a copy of Cary's Map spread out on the carpet, while he turned to his valuable collection of the works of old authors; and thus did we proceed in marking the names to be introduced on the map, and at the same time each of us marking the corresponding name in his own copy of Cary's ' Index Villaris.'

" In those gleams of new light thrown on the dark pages of our history, we had many pleasant discussions; for, in Wales, I had picked up the meaning of many British names of places; and the jocular barrister would sometimes laugh most heartily at some of my explanations."

While Mr. Cary was engraving the map, business in surveying some of the collieries of the Forest of Dean, in completing the drainage of a large tract of marshes at Minsmere near Dunwich, in Suffolk, and in draining and improving some property of Mr. Arbuthnot in Northamptonshire, and other engagements in Kent, Wiltshire, and Somersetshire, contributed to the completion of his great work. Of these professional labours it is only necessary to notice the bold and remarkable plan of drainage by which the Minsmere Marshes were enabled effectually to discharge their super-abundant water into the open sea through a gravelly beach. This was accomplished (in 1814) by bringing to one point close to the beach the great discharging drains, and uniting their currents in a hexagonal channel or well, from which a large cast-iron pipe was laid right through and deeply buried in the pebbles and sand, which were the natural barrier against the sea. This pipe was of course stopped by the sea rising at every tide, and often buried at its mouth by the accumulated load of "shingle" and sand thrown up by the waves. But on the turn of the tide, the pent-up inland wa-ter gradually made its appearance by the opening of a pair of doors in the upper part of the tube ; then a second, and afterwards a third pair of valves was opened by the force of the water, which soon left the marsh ditches and swept away the accumulated pebbles from the aperture of the tube. In process of time such a tube may be expected to become in parts converted to plumbago, but the iron removed as oxide may cement the surrounding pebbles into a mass more du-rable than the original tube.

According to a custom which it is pleasing to record, the observations he made at Wold Farm in Northamptonshire were communicated to his much-loved friend at Farley in the following letter, which also preserves some other in-teresting particulars of their confidential intercourse :—

" London, Feb. 11, 1813.

" DEAR RICHARDSON,—I have lately been into Northamp-tonshire on business for the gentleman who franks this. The estate is near Thrapston, on the crop of the cornbrash, and a part of the strata above and below it. There are some novelties in the appearance, which leads me to suspect the accuracy of our knowledge of the cornbrash. They had a rock opened for building-stone about six feet thick, covered with blue and yellowish tenacious clays containing many fossils, but this did not satisfy me; and about twenty feet higher I found another rock, by sinking and boring, which contains an immense quantity of water. This I traced for a quarter of a mile, considerably covered with clay, and in other parts so completely covered with alluvial matter (from the chalk hills) as not to be found for some miles.

" The rock which I discovered is clearly the upper part of the cornbrash in small thinnish stones, hard and grayish within, and rough on the surface. The most remarkable fossils were the large-ribbed oyster (*Ostrea marshii*), somewhat like our hogs-ear oysters at Combe.

" I have the satisfaction to inform you that the large plates of my Map of Strata are nearly finished. Sowerby is engraving some of my fossils, and I have had a much better offer from a bookseller to publish the first quarto volume than I expected.

" As the season for a revisal of the locality of indigenous plants is just approaching, I hope you will not forget to make a complete list of them on each stratum. This, with your able assistance, would form a most interesting chapter, and would serve to draw the attention of many to the subject of strata who probably might otherwise never think of it. Hoping Mrs. R. and friends are well, I remain,

"Yours truly,   Wm. Smith."
" Rev. B. Richardson, Farley."

In 1814 some portions of the Map were completely coloured, particularly four sheets of the vicinity of Bath, perhaps the most varied and beautiful sheets that have ever appeared in geological colours; and Mr. Smith was allowed an opportunity of explaining them, or rather of delivering a lecture upon them, before the President (Lord Hardwicke) and other members of the Board of Agriculture, *at the time that the allied sovereigns were entering London*. How the auditors kept their places under this excitement is not known, but one of them, Benjamin Hall, Esq., M.P. for

Glamorganshire, son-in-law of Mr. Crawshay, then deceased, a gentleman who took much interest in the proceedings of the Royal Institution, requested a second and private interview. He then reminded Mr. Smith of the strong friendly interest which had always been felt in the progress of his researches by Mr. Crawshay, who had subscribed £100 toward the publication of them, and had paid half of this sum; he expressed his own desire to complete Mr. Crawshay's wishes, and to accompany the payment by an additional subscription on his own part. Sir Joseph Banks, Mr. Coke, the Duke of Bedford, Lord Hardwicke, and some other eminent individuals, also contributed to soften the dire aspect of utter poverty which now in the very crisis of scientific success began to frown upon the author of the 'Map of the Strata.'

In prosecuting engagements of a professional nature in 1814, Mr. Smith was much in Cheshire, surveying coalmines, &c.; in the vicinity of Lynn (inspecting sea-banks); in Suffolk, Kent, the Forest of Dean and Somersetshire; but the greater part of the year was occupied in the hard work of completing the 'Map of the Strata.'

It was very trying work for the publisher as well as the author. The basis of the Map, as already explained, was in many respects peculiar; the colouring of it was more so. Instead of the *flat colouring* ending in narrow defined edges usually employed for maps, Mr. Smith introduced a peculiar style of *full tints* for the edges of the strata, *softened* into the paler tint employed for the remainder of the area which they occupied on the surface. This new style of colouring gave a picturesque effect to the map, but required more than usual skill and patience to be correctly executed, and occasioned great trouble in examining the copies. The colouring of the map was thus rendered more expensive than had been anticipated, and notwithstanding Mr. Cary paid liberally for the labour, it was not always at first properly performed.

At length the difficulties inseparable from such a task were so far overcome, and this enormous labour was so far completed, that a coloured map of the strata of England and Wales was submitted to the consideration of the Society of Arts, supported by various testimonials of its general accuracy and value, in April and May, 1815*. The result was the award of the premium of £50, which had been in vain offered for very many years for a work of this description—a reward which Mr. Smith might have claimed long ago, had not an honest desire to produce his work complete withheld the attempt. The Map was published on the 1st of August, 1815, dedicated to Sir Joseph Banks, and from that hour the fame of its author as a great original discoverer in English geology was secured. Would that this epoch of his revived and enlarged reputation had also been the dawn of more prosperous fortunes, or that, satisfied with the degree in which he had accomplished his gigantic task, he had left to others the completion of his work, and devoted himself for a time to even the humblest of those professional labours by which he had been at least supported through oppressive difficulties, and by which he must have already grown comparatively rich but for the incessant drain of money in following up discoveries which no living man could reasonably hope to complete! If this be censured as the scholium of a feebler mind and less fervid temperament than that which led Mr. Smith through his mighty enterprise, some allowance may be made for the feeling of the writer, who in this year, at too early an age, began to enter the shadow of those calamities in which his revered relative was plunged.

Science, indeed, is a mistress whose golden smiles are not often lavished on poor and enthusiastic suitors. Even in these days, when the Pension List has been opened to literature and science, the rewards are not measured by age,

* "May 14, 1815.—Began at nine in the morning with an artist to colour for me the *first* printed copy of the ' Map of the Strata ' on canvas. May 22, 1815.—Finished colouring the *first* ' Map of the Strata ' on canvas. May 23, 1815.—Attended a meeting of the Board of Agriculture with the *first* finished copy on canvas of my ' Map of the Strata.' "—MS. *Diary.*

genius or poverty; the march of knowledge amongst the community is encouraged by the grants to individuals, but leaders and veterans in the ranks of knowledge can scarcely be thought *rewarded* by quarterly grants of £12 10*s.* and £25 for the short term of their natural lives. In 1815 the name of "scientific pensions" was not coined, but there were not wanting persons of station, knowledge and humanity, who, esteeming Mr. Smith and admiring his solitary and ceaseless industry, exerted themselves to save him from the sad fate which seemed to await him.

The time for a strenuous exertion was indeed come. Geology had kept him poor by consuming all his professional gains; the neglect of his employers too often left these unpaid : in such a condition one unfortunate step was ruin, and that step was made. On the property which he had purchased near Bath, and which he had greatly improved, he was tempted to lay a railway for bringing the freestone of Combe down to the Coal Canal, to open new quarries of this stone, and to establish new machinery for cutting and shaping it for buildings. The project, which looked well at first, failed utterly by the unexpected deficiency of the stone, on whose good quality the whole success depended. The abandonment of this cherished scheme was followed by the compulsory sale of the still more cherished property, a load of debt remained to be discharged, and the miserable effects fell heavily on others besides himself.

Such things are common in the lives of men, but they are not often encountered by so resolved and patient a spirit as that of Mr. Smith. One who saw the struggle may boldly say this, because there can be no other motive for mentioning private and personal griefs but to show forth the character of the mind which could firmly bear and overcome them. As a means of reducing his difficulties he proposed to sell that geological collection which had been so much prized, and through the assistance of some friends a communication was opened with the Treasury. Two gentlemen being deputed to examine the collection, reported favourably, and their Lordships were pleased to authorise the purchase, in order that the specimens might be fitted up in the British Museum. There was also some defined notion of engaging Mr. Smith's services at the Museum to take charge of and explain the geological principles which this collection was intended to illustrate; but this project came to nothing. The sum of money granted for the purchase was £500 (January 1816). By a later order the further sum of £100 was allowed for an additional series obtained from North Wilts, Essex, and other localities, to complete the series (February 1818); and £100 was granted to compensate for trouble of arrangement, making catalogues, &c. The whole sum was £700; the number of species supposed to be 693, and of specimens 2657. The authorities at the Museum assigned first one and then another apartment to the collection, and it was arranged according to the peculiar views of the framer; but the Museum then possessed no department of "palæontology" or "geology," and there was a difficulty in making these specimens, *which were arranged on sloping shelves, to represent the strata,* a part of the series of objects open to the public. The present state of this "the first stratigraphical collection" ever made, is unknown.

In 1817, a portion of the descriptive catalogue of the collection sent to the British Museum was published under the title of "*Stratigraphical System of Organized Fossils,*" in which the fossils collected from the British strata, as far down as the "marlstone," are mentioned, with careful notices of the localities. Curious coloured tables are added, the first of the kind ever published, showing the geological distribution of particular groups. In this year also was issued the first number of another work, entitled "*Strata Identified by Organized Fossils,*" consisting of numerous figures of fossils engraved by Sowerby, and printed on paper to correspond in some degree with the natural hue of the strata. This remarkable work reached its fourth number (only seven were proposed). The publication of it was undertaken by Mr. Sowerby, in consequence of an arrangement by which William Lowndes, Esq., of the Tax-office, a very strenuous and judicious friend of Mr. Smith, advanced £50 to pay for the cost of the first number. The expense of this work left to the author very little chance of profit. Mr. Sowerby estimated the cost of each number at £50; the gross sale price, supposing the whole (250 copies) to be sold, would yield £93 15s., from which the expenses of publication, bookseller's charges, &c., were to be deducted.

Now also began the publication, by Mr. Cary, of a large series of geological sections and county maps, coloured upon the same system as the great Map of England and Wales, and carrying out to greater minuteness the delineation of the boundaries and areas of strata. This series extended to twenty-one English counties, including in four sheets a large and valuable map of Yorkshire.

The practical applications of new discoveries in science sometimes extend faster and further than the knowledge of the general principles on which they depend. A correct and generalized knowledge of the nature, succession, and position of strata, in any given region, is a necessary antecedent to correct general practice in draining land and in obtaining supplies of water in new situations, but special cases occur in which drainage can be accomplished and water obtained by the help of experience and analogy. This has been fully shown by the common process of sinking Artesian wells, which are successfully repeated in many districts where an accidental or random experiment had succeeded before. But to direct the attempt for such wells in new situations requires the aid of *science*, as distinguished from *experience*.

A remarkable exemplification of these views occurred in 1816, in the vale between Swindon and Wotton Basset, on the line of the Wilts and Berks Canal. The Canal Company had been much inconvenienced by the scarcity of water to feed their own summit level and supply the down lockage of the North Wilts Canal, and being much restricted in the use of the natural streams, they commenced, without geological advice, a well, or pit, to obtain water from below the surface. The pit reached the depth of about 40 yards, and a boring had been continued about 40 yards deeper, without success; Mr. Smith was then appealed to for advice in respect of further proceedings. The investigation which now followed is both instructive and monitory. Part of it may be given in Mr. Smith's own words, as a specimen of his reports on business.

"Swindon, April 13, 1816·

"In order to form a correct opinion of the success of the experiment for water now going on by the side of the Wilts and Berks Canal, I have particularly examined the nature of the earth sunk and bored through, and endeavoured by local observations to ascertain the extent of the stratum now penetrating, and the nature of the rock be· neath, which is expected to produce the supply of water required.

"The stratum in which the pit has been sunk to the depth of 46 yards, and bored into near 50 yards deeper, is chiefly a tenacious clay, containing at certain depths layers of the *Septarium* or *Ludus Helmontii*, very similar to those from which Parker's Roman cement is made.

"These stony nodules the sinkers have called rock, but no regular rock has yet been found, nor is there any hope of finding any until the whole stratum of clay is penetrated. The depth of such a perforation can only be judged of by similar experiments for coal and water in various places along the course of this extensive stratum. Here the depth was, of course, expected to be great, from the known depth of several deep wells in the neighbourhood, all of which produced water which ascended to their tops; and the deepest and nearest to this experiment having done so and continued to overflow ever since it was sunk, afforded data sufficient for such a proceeding. Besides the water found at Mr. King's of Mannington Farm, I find that water has been obtained at another farm of his, and at Costar and Whitefield, along the course of the same clay-ridge which extends to Wotton Basset; and that at three of these wells, like that of Mr. Edwards's well (at Even Swindon), the water is of a mineral quality. All of them, I am informed, have a copious supply of water, and stand full to the surface, or

G

nearly so, which proves the original source or head of the
water to be on high ground. This my extensive and long-
continued observations on the strata led me to expect; and
the order of the strata is that which I have always thought
it to be, a thick stratum of clay overlying the coral rag and
upper oolite rocks, which crop out or appear on the surface
at Wotton Basset.

" To apply this general knowledge of the strata to the situation in question, I have particularly examined the outcrops
and extent of surface occupied by that stratum of limestone,
as much of the success of the present experiment must depend upon the extent and cavities for water in the stratum
which underlies the clay.

" The surface of this rock about Wotton Basset is very
narrow and interlayered with clay, but between Lidyard and
Purton much more extensive and absorbent, terminating
about the latter place on the borders of the river Roy, in long
narrow ridges sloping to the east, with a partial declination
toward the south or south-west, which has a tendency (with
the other outliers of the rock) to form a basin or trough,
whose deepest part is near the present boring-hole, and
consequently the greatest quantity of water which the rock
will produce may be there expected. Thus far is the
theory of finding water at that situation correct, but there
may be some practical objections to the quantity it will produce :—

" 1st. The top part of the rock is covered, and frequently
interlayered with clay.

" 2nd. The whole rock is not above 20 or 30 feet thick.

" 3rd. It has but few open joints, and those not very
large.

" Consequently the faults in the declination of these strata
to the east or south-east may frequently interrupt the general descent of water to one point in the deep, and occasion
considerable partial discharges of the water absorbed, as I
found in the little valley below Purton.

" Yet this copious discharge from one or two hundred
acres of land serves rather to prove how much the re-

mainder of the rock, which has no visible discharge, may be expected to produce.

" Water will most probably soon be found, which may be expected to rise to the surface, but with such a head upon it, the discharge will be slow unless it be assisted by machinery, and the natural apertures at the bottom of the pit enlarged by some tunnelling into the rock; all of which is very practicable. Should there be any variation in the quantity of water, it is generally, under such circumstances, most abundant in summer.   " WM. SMITH."

These predictions were verified. The clay was passed, the coral rag and oolite were entered, and in the upper part, consisting of a few feet of compact rock interlaminated with thin clays, only a moderate addition of water was experienced. The workmen proceeded to sink the pit to this rock and to bore deeper, and, as Mr. Smith foretold, the quantity of water augmented continually, till the well altogether reached 87 yards in depth, the lowest 5 yards being bored through rock. The supply was found to be considerable; the uprushing water filled 15 or 18 feet of the well with sand brought from below the limestone rock, and the assistance of a steam-engine was required to prosecute the still incomplete experiment. The water rose in the well, in December 1817, through a bore-hole of 3 inches only, 23 yards 1 foot in 26 hours, notwithstanding the most active efforts of the horses employed, and the ultimate success of the experiment appeared probable. The geological investigation had fully proved the continuity of the oolitic rock, and ascertained that it was sufficiently expanded on the surface to gather a good supply of water from rain. The access of the water treasured in its subterranean area to the base of the well seemed to be free, and Mr. Smith believed that only one thing remained to be done, viz. to drive headings or water-levels *below the rock,* so as to command the streams flowing in the fissures through a sufficient breadth of ground. The following was his plan.

" The first sudden rise of water into the well through

such a small aperture, and the subsequent rush of water through the sand above referred to, shows that the water comes freely to this hole; but to obtain a supply sufficient for the regular work of a 50-horse power engine will require headings driven in the level course of the stratum which produces the water, unless the natural working of the water through the sand, with an enlarged aperture, should of itself make a sufficient aperture. Headings may be driven in this loose sand, under such a rock roof, to any extent and at a moderate price. Such subterraneous cavities or headings, if made capacious enough to hold from 10 to 20 locks full, would serve as regulators between the flowing of the spring and the inequalities of consumption."

To what extent this was performed does not appear in the documents which Mr. Smith has preserved. In 1820, however, the supply of water was found to be very limited, and the experiment was abandoned. The section on page 85 correctly represents the principal facts of this curious case, which may be thus summed up:— 1. The well was sunk and bored about 80 yards in clay, without water, except a small chalybeate spring at 14 yards. 2. Mr. Smith was then consulted, and upon general principles declared that no great quantity of water would be found till a jointed rock was pierced. 3. That rock he declared would be found by continuing the experiment, and would yield water. 4. This was found correct, but either from an insufficient subterranean extension of the works, and from the comparative thinness of the oolitic rock (only 21 feet in this well), or from some other peculiarity of the ground, the supply failed. It arose perhaps from the circumstance that only a narrow breadth of the rock, inclosed between small faults, conveyed its watery stores to this pit and was soon exhausted; it is certain, from the height to which the water rose, that it was pressed upon by a column whose effective weight reached nearly to the surface; and if the headings alluded to by Mr. Smith had been boldly driven right and left, as in a colliery, it is probable that a large and constant stream might have been found and the experiment have been finally

SECTION OF STRATA,—NORTH WILTS.

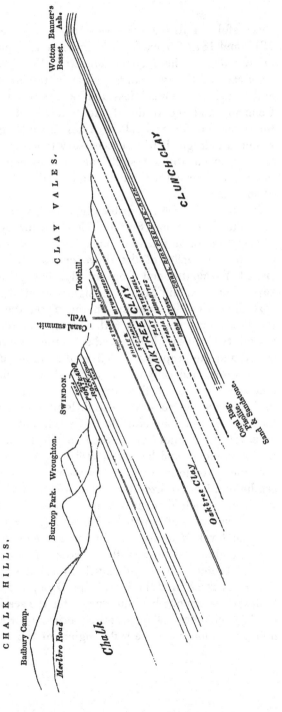

CHALK HILLS.

Badbury Camp.

Burdrop Park. Wroughton.

SWINDON.

Canal summit. Well.

CLAY VALES.

Toothill.

Wotton Banner's Basset. Ash.

Marlbro Road

Chalk

PORTLAND ROCK. SAND.

Oaktree Clay

THIN STONE

SHALE B.

SEPTARIA

OAKTREE CLAY

OYSTER SHELL

AMMONITES

SEPTARIA FLAT

IRON STONE

BITUMENIZED WOOD

WHITES

CORAL RAG PISOLITE OOLITE

Coral Rag; Pisolite.

Sand & Sandstone.

CLUNCH CLAY

successful. This reasoning was employed by Mr. Smith in 1811 and 1812, to explain the numerous separate springs on the line of the Kennet and Avon Canal, which issued from each of the two oolitic rocks in the hills near Bath, and occasioned so much loss and inconvenience to the Canal Company and the trade of the country. In experiments for water on the principle of Artesian wells, if the supply required is large, it will be necessary to avoid alike a hasty commencement without sufficient local examination, and a hasty abandonment without a complete trial of all the means of success.

In 1817, Mr. Smith's professional engagements carried him into Norfolk and Suffolk, where he surveyed and prepared plans for an intended line of river and canal navigation down the valley of the Waveney from Diss to Bungay, visited Yarmouth, and from an elevated point there surveyed, with Captain Manby, a mighty sheet of flood waters spreading over the marshes on the Yare, the Thurn and the Waveney. He saw it with regret, and remarked that though twelve years had passed since the sea-breaches were closed and the proprietors, delivered from the dread of the outer water, had only to contend with the upland streams and the swelling of the rivers, the improvement of the drainage seemed to have been little prosecuted. Perhaps had he remained in Norfolk this would not have been the case, for the confidence reposed in his skill and resources was unbounded, and he was gratified by the assurance that " by the introduction of water-meadows and expulsion of the sea he had immortalized his name in Norfolk." Mr. Smith had been consulted at various times concerning the drainage of certain tracts of marsh-land near Lynn, which were below the level of high-water in the sea and in the rivers. This neighbourhood was the theatre of a mighty work, the celebrated and long-delayed Eaubrink Cut, by which the great river of " the fens," the Ouse, was to be forced to desert its sinuous bed and encounter the sea in a straight artificial channel. This grand experiment for assisting the difficult drainage of the wide region of the fens was the subject of much controversy among professional men; for it was argued that if the new cut would permit the drainage water of the uplands to pass more rapidly to the sea at Lynn Deeps, along its straighter and steeper channel, it would also admit the tide more easily and permit it to flow further and higher up the river, and thus increase the difficulty of the drainage of the lands far up the country, though it might facilitate the discharge of water from the lands nearer the outlet*.

It is a remarkable fact that the level of the marshes embanked at successive times from the sea rises continually; the outer marshlands being higher than the inner fens, and the unembanked foreshores higher than either. Such at least was the case in 1817, when Mr. Smith inspected a sea-bank of Lord William Bentinck's, then in much danger from the powerful tides of Lynn Deeps. Nor is this at all a solitary case. It is in fact one of most frequent, even ordinary occurrence, on other parts of the British shores, and may be witnessed along most of our great tide rivers: for example, the Severn, about Fretherne. It is however more remarkable in the great fen country, on account of the distinct succession of at least three levels, as in the annexed sectional sketch, where the innermost lands (fenlands with interspersed lakes) (1) are the lowest and of least value, the next district toward the sea (called "marsh land") (2) a few feet higher, and of very great value (£3 per acre), and the margin of the sea-coast or foreshore, (3) highest of all, growing daily in breadth and height, and tempting the proprietor to dare the ha-

* This was apparently Mr. Smith's opinion.

zardous encounter with the sea. In this country the means
of embankment are silt dug from the marshes, grassed over
or protected by timber, jetties and other constructions; but
this process was too unlike the work of Nature to please
the inquirer who had stopped the dreaded sea-breaches with
sand and pebbles.

From the eastern coast he was called into Yorkshire to
consider of a plan for a new canal between the river Aire at
Knottingley and the river Dun at Doncaster, with a branch
down the river Went. On this matter he was employed at
frequent intervals till 1819, when the bill was brought into
Parliament, and defeated by the strenuous opposition of the
Aire and Calder Navigation Company. In the course of
these surveys he became perfectly acquainted with the series
of limestones and gypseous clays and sandstones of the magnesian limestone and rothe-todte-liegende, and made several
visits to the West Riding coal tract. The oolitic moorlands
and lias coast of Whitby and Scarborough were included in
some journeys of this period (1817); and a very interesting
excursion was made in the early part of 1819 to the ancient
mines of Swaledale ('Auld Gang Mines'), where the agents
vied with each other in contributing exact sections of the
limestone series and plans of the veins and faults, and conducted him through several of the remarkable works. The
geological results of all these journeys in Yorkshire were
coloured on the large County Survey of Jeffreys, which had
been employed for the same purpose in 1803 and subsequent
years, while examining the country about Spofforth, Mid-
dleham, and Peirse Bridge, and were after other expeditions
in 1820 and 1821, published on the four-sheet map of Cary.
A variety of engagements of like nature led Mr. Smith in
many directions from London, and enabled him to complete from time to time those valuable county maps and sections already mentioned. The Forest of Dean was frequently visited in 1817, 1818 and 1819, and there was hardly
a colliery in that singular and picturesque property of the
Crown which he had not examined in detail and illustrated
by original plans and sections One of these plans was of

much interest, for it exhibited across a part of the forest a
tract of "dead ground," crossing in a winding course one of
the coal-beds. This dead ground, called the "Great Horse,"
is entirely unconnected with faults, and is in fact merely a
subterranean tract in which coal would, according to all the
usual laws, be found of the usual thickness, but in which,
from some peculiarity of the original deposit, no coal occurs.
If the coal-bed be thought to have been formed from a peat
bog, this "horse" may be regarded as an *ancient water-
channel* in such a bog: and perhaps that is a conjecture
very near to the truth.

In the early part of 1818, a considerable part of Monmouthshire was the subject of special geological surveys,
which enabled Mr. Smith to draw a section from the centre
of the Forest of Dean across the limestone and old red sandstone by Monmouth and Ragland to the same strata near
Pontypool, Risca, and Mynnyddysllwyn, in the midst of
the coal-basin of South Wales. Mr. Smith was perfectly
aware of the great analogy, and perhaps even identity of
the series of coal-beds in the Forest of Dean with those
of Glamorganshire, the ironstone courses being wholly de-
ficient; he was fully in possession, by means of accurate
surveys about Monmouth and near Laugharn, of the detailed sections of the old red sandstone ; but he was only
very slightly acquainted with the nature of the Silurian district near Usk, which on his map is indicated by patches of
limestone in the midst of the "red rhab."

To the gratification which, as an engineer, Mr. Smith had
often experienced where his art enabled him to direct and
to triumph over the turbulent powers of nature, was now
added the sweeter and nobler reward due to a man of science.
In this year (1818) his claims as a great discoverer in English geology were fairly and fully advocated by one whom
honest inquiry had satisfied of their truth, and filled with a
generous desire to redeem from neglect, not "Strata Smith"
alone, but many earlier and meritorious names which were
fast disappearing from the roll of English fame. Dr. Fitton's 'Notes on the progress of English Geology' (which

first appeared in the Edinburgh Review in 1818) are too well known and valued by the geological world to require commendation here; but it is necessary to Mr. Smith's personal history to remark, that it was the favourable light in which Dr. Fitton's unsolicited kindness placed Mr. Smith's name and labours, which maintained and augmented public interest in his long and solitary labours, stimulated the members of the Geological Society of London to an impartial estimate of the claims of one who was not their associate, and procured for him in 1831 the award of the first Wollaston Medal, and in 1832 the grant of a pension from the Crown of one hundred pounds a-year. To the well-timed and generous effort of Dr. Fitton Mr. Smith owed, and was glad to acknowledge it, much of the comfort and consideration which relieved the otherwise heavy gloom of his declining years.

In the winter of 1818–19, Mr. Smith revisited, after an absence of ten years, his native village, re-examined the unforgotten localities where in childhood his " pundibs " and " poundstones " were gathered, and collected "marlstone" fossils from an excavation at Churchill Mill, nearly at the same points where he had noticed them in 1787. In one whose life had been one long wandering, and who had earned for himself an immortal name, this return to the haunts of his childhood and the simplicity of village occupations, must have excited many interesting reflections. He had sold his patrimony, and what had been the modest dwelling of his ancestors for 200 years, a house such as many *were* in that village, literally

"Congestum cæspite culmen ;"

he had disbursed in travelling for what he deemed a public object all that he had received, and become answerable for all that he had earned; while one of his two brothers, quietly prosecuting trade in his native village, had grown a rich and prosperous man. Many changes had come over that village. One of the largest and most honoured elms had fallen; the great common field was but a name; it could no longer be said, as previous to the inclosure, in 1787, might have been at least poetically said, that

"Every rood of ground maintain'd its man ;"

the yeomanry had sold their " yard lands," and been transformed into tenantry renting the broad acres of the squire.

"Times were alter'd ;"

there was no longer a treasurer for the "Whitsun-ale," but the "wake" was still a scene of merriment, to which, among other idlers, the gypsies still gathered from "the Forest" of Whichwood; and Mr. Smith yet found among the old inhabitants some who could remember the digging of Sarsden pond, with its "golden" stones (iron pyrites), and the ornamental planting of Daylesford by "Governor Hastings," and laugh with him over the marvellous tales of "horses having run their feet off" in dragging the "fly coaches" on the Oxford road at a pace inconceivable to the slow Saxons of the "Cotteswolde Hilles."

In the autumn of 1819 Mr. Smith gave up his house in London, after fifteen years' occupation, and was compelled to submit to the sale of his furniture, collections and books, preserving in fact only his papers, maps, sections and other drawings, through the kindness of a most faithful friend. While this happened he was in Yorkshire busily engaged, apparently oblivious, perhaps sternly regardless, of what seemed to others an insupportable misfortune. He deemed it an inevitable corollary to his irretrievable losses in the unlucky speculation already mentioned near Bath, and armed himself with what seemed more than fortitude to meet it.

One more used to monetary arrangements would have foreseen and averted this occurrence ; but on the abstract geologist the blow fell with stunning effect. He surrendered with deep regret his interest in the much-loved and really valuable little property near Bath, quitted London, and consented to have no home. From this time for seven years he became a wanderer in the North of England, rarely

visiting London except when drawn thither by the professional engagements which still, even in his loneliest retire-
ments, were pressed upon him, and yielded him an irregular,
contracted and fluctuating income.

In the winter of 1819–20, Mr. Smith, having perhaps
more than usual leisure, undertook to walk from Lincolnshire into Oxfordshire. On this, as on almost every journey
for the last three years, the compiler of this memoir was his
glad companion, " haud passibus æquis," and, according to
an established custom on all such tours, he was employed
in sketching parts of the road, and noticing on maps the
geological features of the country. The object proposed
was to pass along a particular line through the counties of
Rutland, Northampton, Bedford and Oxford, but the ultimate destination was Swindon in Wiltshire. Leaving the
great road at Colsterworth, with some reflections on the
birthplace of Newton*, we crossed in a day's easy walk the
little county of Rutland, its hills of oolite and sand, its
slopes of upper lias, and its valleys often showing marlstone, and reached the obscure village of Gretton, on the
edge of Rockingham Forest. Whatever may now be the
accommodations at this village, they were very wretched in
1819 (December), but the odd stories of supernatural beings
and incredible frights which were narrated by the villagers
assembled at the little inn, greatly amused Mr. Smith, and
reminded him of exactly parallel tales which circulated
round Whichwood Forest in his boyhood.

The next morning we walked to Kettering, noticing on

* " Sir Isaac Newton was a promoter of geological investigation ; but
he, like others of the day, looked to things at a distance rather than at
home. It was not an object for a telescope. Newton's own fields, or at
least those he must have often walked over, are literally strewed with fossils in a manner which I never saw in any other soil, lying thereon like
new-sown seeds of oats, and so numerous are they (where I observed
them), that in the moist state of that tenacious soil the great philosopher
may have scraped them (unobserved) from his shoes by hundreds. It was
this which, on my receiving the Wollaston prize, induced me to say that
' had Newton condescended to look on the ground he must have been a
geologist.' "—MS. dated May 20, 1839.

the road the peculiar characters of the Northamptonshire oolite. In this walk Mr. Smith had somehow sprained or overfatigued himself, and he chose to proceed to Wellingborough in a chaise. From this point, situated on sand of the oolitic series, we resumed our geological proceedings on foot, and passing by Irchester, Woolaston and Boziate, traversed in the next hills the oolite, the forest marble, the cornbrash, and an outlier of Kelloway's rock. The road up Boziate Hill was mantled with fossiliferous stone, some of which obtained from the hill-top was believed to be Kelloway's rock, and was found to contain *Ammonites sublævis* and other fossils. A fine specimen of this ammonite was here laid by *a particular tree on the road side*, as it was large and inconvenient for the pocket, according to a custom often observed by Mr. Smith, whose memory for localities was so exact, that he has often, after many years, gone direct to some hoard of this nature to recover his fossils. This road, however, over Boziate Hill, he was not to travel again.

From Olney to Buckingham the route was performed in chaise. The stone dug here in clay attracted much attention, and Mr. Smith doubted whether to rank it as forest marble or cornbrash. We now crossed the oolitic country to Aynhoe, celebrated for its fossils, on foot; next day continued the walk to Deddington, Chapelhouse and Churchill, and after a few days walked to Burford, and then travelled in the ordinary way to Swindon, Oxford and London. In passing through Oxford, Mr. Smith, for the first time in his life, had the pleasure of seeing Professor Buckland, at the house of Mr. Bliss, the bookseller, with whom he walked over Shotover Hill, on his way toward London.

This little tour is thus briefly narrated, because it appears in all respects a fair example of the usual way in which Mr. Smith explored the country, walking when the object he had in view required this mode of examination, travelling as fast as possible in all other cases, but always recording in note-books or on maps the observations he made. The subjoined section is taken from one of the note-books :—

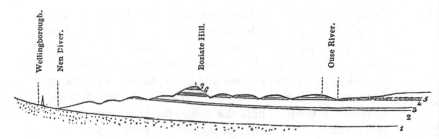

1. Brown sandstone of Wellingborough . . . .
2. Oolite and clays.
3. Forest marble.
4. Clay.
5. Cornbrash.
6. Clay.
7. Kelloway's rock . . . .

Early in 1820 Mr. Smith was employed in Leicestershire, but the greater part of the year was passed in Yorkshire, about Ferrybridge, Whitby, and Scarborough. While staying at the romantic and delightful town just named (in hopes to soothe the mental aberration of his wife, which became very manifest in this year), he imbibed for it a partiality which augmented with further knowledge. Mr. Dunn, one of the most affectionate and highly esteemed of his friends, remembers that even at this period it was in contemplation to found a museum at Scarborough, and that Mr. Smith attended a private meeting of a few inhabitants for the purpose, but the project was not urged into effect till after some years had elapsed, when the same individuals undertook the task of establishing the Scarborough Philosophical Society with better omens and stronger assurances of support.

Early in 1821 the writer walked through the eastern parts of Yorkshire and rejoined Mr. Smith at Doncaster, and from this point accompanied him in a walking excursion through the coal district of the West Riding, passing by Bamborough, Houghton, Cudworth, Shafton, Wakefield, Ardsley, Horbury, Thornhill (the Rectory of the Rev. John Michell, the celebrated Woodwardian professor), Flockton, Bretton, Haigh Bridge, Silkstone, Stainborough Inn, Wentworth Castle, Tankersley Park, Wentworth Park,

Rawmarsh and Conisborough. In this excursion particular attention was given to determine the true general order of the coal beds, ironstone courses, and characteristic rocks, and the result is seen in a comprehensive section on the Yorkshire map, to which nothing similar had ever been attempted in this country, perhaps in Europe. From the notes and sketches made on this journey, the first page is taken to show the kind of information which it was proposed to gather, besides the lines of outcrop of the strata marked on the map.

"4th April, 1821.—Walked westward by Barlby and Warmsworth to Sprotbrough Ferry, and thence by Cadeby over the verge of the magnesian limestone there. Cadeby is so very ill supplied with water, that those who hold the pump-handle must frequently wait some time till the niggard stream furnishes its scanty supply. There is, however, a pool in the place, and some springs appear above the edge

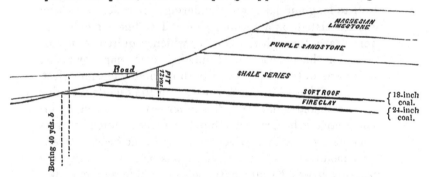

of the hill which, as at Micklebring, is kept by the purple sandstone. Several yew-trees mark its course. Before we arrive at the mill on the Dearn, some clay ground and dark shades in the soil indicate the outcrop of the uppermost workable vein of coal. Melton Hall, to the right on the edge of the limestone, is very conspicuous. We now ascend a bold bank, and at the junction of the Barmborough and Melton road, wind round a sudden swell of limestone and sandstone, whose outcropping beds form around the face a series of natural entrenchments clothed with wood. Beyond

this hill the ridge recedes from our road and leaves a broad
vacancy, wherein we descry a coal-pit. To our left is
Barmborough Grange, standing on detached sandstone.

" The boring marked *b*, now prosecuting with the hope
of piercing the two coals found at Goldthorpe and Thurns-
coe. The upper, at Goldthorpe, 70 yards deep, thin and
not desirable; the lower 105 yards deep, 5 feet thick*.
Solid sinkings, soft floor, but immediately beneath hard
sinkings repeated."

To assist Mr. Smith's researches the author made a variety of other journeys from Doncaster, as to Roche Abbey,
Gringley-on-the-Hill, and Gainsborough. At length the
materials for completing the geological map of Yorkshire
were deemed sufficient, and this remarkable work was published in the summer of 1821, in four sheets.

The desire to finish others of these interesting county
maps, led Mr. Smith to devote the whole of the remainder of
1821 to long and laborious wanderings, in which the writer
was associated, on a peculiar plan. Two lines of operation
were drawn through the country which required to be surveyed for the purpose of completing such maps, or rather
such parts of the maps as had been inevitably left imperfect.
On one of these Mr. Smith moved with the due deliberation
of a commander-in-chief; the other was traversed by his
more active subaltern, who found the means often to cross
from his own parallel to report progress at head-quarters.
This mode of " strata hunting" was not necessarily expensive; it was besides extremely agreeable and effective,
and was faithfully executed in peregrinations which lasted
six months, and permitted one of the parties to walk over

* On this map, for the first time, an attempt, by no means unsuccessful,
was made to divide and colour the great coal district into groups of rocks
and shales; the course of the " Pontefract Rock," " lower new red sandstone," or " rothe liegende" was traced out, and the magnesian limestone
divided into its component members. The first copies of this map have
one grand error in the north-eastern moorlands, which are represented to
consist of oolitic strata not lower than the " Clunch (Oxford) clay,"
whereas the great subjacent shale is lias. This, however, Mr. Smith
quickly discovered and rectified.

2000 miles of ground, and preserve memoranda of almost every mile along that line.

The first " fytte" of this tour led from Doncaster through Lincolnshire, by Gainsborough and Harpswell Hill to Spittle, Market Raisin and Lincoln. A good section of strata had lately been exposed in the road cutting up Harpswell Hill, and as this has perhaps not been elsewhere recorded, it is here subjoined, as an example of the ordinary structure of the edge of the oolitic ranges of Lincolnshire:—

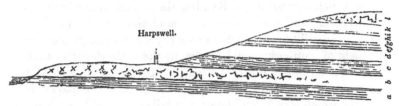

Harpswell.

l. White limestone (oolite), 30 feet.	*e.* Clay parting.
k. Whitish clay and sand, 12 feet.	*d.* Brown sandstones and shells, 30 feet.
i. White sand, 3 feet.	
h. Sandy, with iron balls, 10 feet.	*c.* Blue clays, 20 feet exposed.
g. White micaceous sand, 3 feet.	*b.* Marlstone series.
f. Brown sands, 4 feet.	*a.* Lias clays.

During a short sojourn at Lincoln the country round the towering cathedral was carefully explored, and the rich quarries of Canwick were rifled of their fossil contents.

One of the pedestrians next passed along the ancient road, here called the " Ramper," to Sleaford, Grantham and Stamford, correcting the outlines of oolite cornbrash and Oxford clay; and, after a very circuitous course to the east, returning to Wandsford, and in the midst of Rockingham Forest, visiting Weldon, a place famous for its stone quarries in the days of Morton, author of the old ' History of Northamptonshire'. The section of Rockingham Hill was found comparable to that of the Lincolnshire " Cliffe," and similarly beneath it in the valleys of Rutland, the marlstone was seen rich with fossils exactly as in the vicinity of Grantham. From the marlstone edge of Tilton-on-the-hill the route proceeded across the lias ground to Leicester. Charnwood

Forest was then crossed, and the contorted limestone hills and coal basins on its border surveyed, the journey ceasing for a few weeks at Nottingham, to allow of recording on maps the information collected.

After examining the circumstances of the neighbouring lias and new red sandstone, the coal-beds about Radford, and the termination of the granular (crystalline) magnesian limestone about Bilborough, we set forth by two lines through Derbyshire, and Yorkshire, and Lancashire, to a new point of union at Kendal, the grand object being to trace the outlines of the strata round and through the Lake district. The eastern line of survey passed across the coalfield of Nottingham and Derby, by the Butterley iron-works to the limestone hill of Critch, thence by the dale of Matlock and the cliffs of Hathersage and Bamford Edge to Peniston, which was found almost naturally paved with the Yorkshire flagstone *in situ*. Almondsbury (Cambodunum?), Elland Edge, the romantic vale of Todmorden, Burnley, Colne, Skipton, all most interesting localities, were eclipsed by the bolder scenes round Settle, Giggleswick, Ingleton, Kirkby Lonsdale, and Kendal. The other line passed further west, and turned north through Preston and Lancaster. From Kendal, one shorter line of operation was followed straight by Grasmere and Thirlmeer to Keswick; another shot off to the east by Sedbergh and the Crooks of Lune, and returned by Orton, Shap Fells, and Longsleddale to Bowness, then encircled the Lake district on the south and west, by Newby Bridge, Ulverston, Dalton, Broughton, Ravenglass and Ennerdale, and turned by Buttermere and Newlands to Keswick.

Keswick was a centre of active exertion, and here the assistance of an excellent geologist and very clever man, Mr. Jonathan Otley, was found to be extremely serviceable, and Mr. Smith was happy to find that his views, gathered from observations specially directed to the point, entirely concurred with those Mr. Otley had already formed and published. The maps of Cumberland, Westmoreland and Lancashire were now prepared, and a more convenient map,

indeed a better one than any of them (Cary's Sheet Map of
the Lakes), was coloured (Sept. 1821), in conformity with
views since universally adopted.

Mr. Smith's attentive eyes embraced every curious object
in nature or art, and found perpetual occupation for the pen
or pencil of his willing assistant. Perhaps the accompanying sketch of a piece of slate from Coniston Fells, dressed
for roofing at Keswick, may be curious enough to deserve
to be remembered. It is seldom that the effect of a slight
displacement of the bands of stratification (*s s′ s″*) is seen so
clearly, and the two spar veins (*v*) which cross in straight
lines both the beds and the fault, add to the instructive cha-

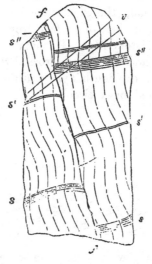

racter of the specimen. The curved lines are edges, more
than usually flexuous and symmetrical, of a scaly structure
lying obliquely to the plane of cleavage.

Before the end of September our pleasant residence at
Keswick was exchanged for another ramble to Aldstone
Moor. But in order to reach this not distant point, Mr.
Smith passed along the side of Bassenthwaite Lake, and
turned by Cockermouth, Wigton, and Hesket Newmarket
to Penrith, while the writer returned to Grasmere, Coniston

Water, Bowness and Kendal, and thence took the road by
Sedbergh, Kirby Stephen, Appleby, and Hartside Fell to
Aldstone Moor. Here we gleaned from experienced persons
the history of the mines and levels, and examined with care,
and recollections of Swaledale, the admirable natural sections and contours of the hills which margin the South Tyne
and the Nent. In the barren region which surrounds this
capital of the " lead-mining " district, we saw no corn but
oats, and very few gardens either beautiful or well-stocked,
yet the market was largely supplied with fruit and vegetables
brought from the fertile vale of Eden. The miners were
liberal and eager purchasers of the fruit; and though their
country, to use the expression of a resident, " grew nought
but grass and lead," the subterranean wealth amply com-
pensated for the absence of surface crops.

On leaving Aldstone Moor, Durham was appointed to be
the rendezvous, and Mr. Smith crossed the high summit of
drainage between the South Tyne and the vale of Derwent,
and thence passed by a circuitous route to Durham, exa-

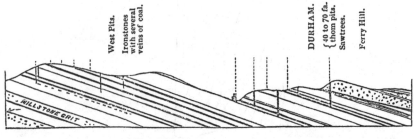

mining the country with care. The writer was deputed to
survey lines stretching northward by Brampton to Gils-
land Wells; eastward along the basaltic cliffs and the
Roman wall to the North Tyne; and thence northward by
the ancient Roman road to the foot of Carter Fell, and the
skirts of the Cheviot beyond Wooler. The return was
made by Belford, Alnwick, Morpeth, Newcastle, Corbridge
and Ebchester to Durham. Here we loitered awhile. As
a general result of the examinations now made in the county
of Durham the foregoing sketch was made.

The expedition now turned southward by two lines through the mountain limestone and millstone grit tracts of Durham and Yorkshire, one passing by Barnard Castle, Reeth and Ripon, the other by Masham, Ripley and Harrogate, to Leeds. Here a friendly welcome awaited Mr. Smith; he was introduced to the most active members of the Philosophical Society then newly formed, and received particular attention from Mr. E. Baines, the late Mr. Atkinson, and Mr. E. S. George. He delivered a lecture or general explanation of his views on the Geology of Yorkshire before the Society, and took much pains to complete his knowledge of the stratification of the vicinity. A map combining these results was at that time carefully drawn by the author of this memoir. After a short visit on business to the lead mines of Swaledale, the journey and the year terminated with a walk from Leeds to Nottingham, where Mr. Smith resumed his old quarters, intending to repose for the winter. But during a short visit to London, an accidental meeting with Colonel Braddyll produced a professional journey to and general mineral survey of some estates of that gentleman in Lancashire, Cumberland and Durham. On this occasion the limestone country round Conishead Priory was investigated with reference to the probability of finding copper ore; and some experiments for this purpose were then in progress. These trials originated in the fact, that in different but not far-removed situations, many lumps of copper ore were discovered in narrow crevices in the limestone, and some in the "pindle" or "diluvial" gravel, which covered the limestone in many places. In a gutter cut across the supposed course of the vein (east and west), "a surprising phænomenon" was noticed, "the surface of the limestone rock far above the sea being polished by the attrition of the gravel resting on it, the direction of the wearing being parallel, or nearly so, to the line of slope of the hill." This was regarded at the time as an effect of "diluvial" action. It may perhaps be referred by modern speculations to the movement of a glacier.

The argillaceous marls of the old red formation were

examined in a neighbouring estate; the slaty rocks, limestone and red iron ore about Dent Hill, Cleator, and Egre-
mont; Whitehaven and the neighbourhood of Hesket Newmarket were visited; but the interest which Mr. Smith felt
in these inquiries was altogether inferior to that with which
he scrutinized the geological position of Colonel Braddyll's
property in Durham, including the village of Haswell.

Here seven hundred acres of poor land, situated on the
magnesian limestone, were held in slight esteem by the
agents, and the absent proprietor had been advised that no
mineral wealth existed below the surface. Mr. Smith at
once saw, by a survey of the neighbouring coal district, that
the limestone was an "unconformable" cover to the coal
strata; these he traced in their courses, estimated the thickness of the limestone, and inferred that the best seams or
veins of coal would be found at an attainable depth at Haswell. He spoke to the neighbouring proprietors, and learned
with surprise that few of them believed their properties to
have any other than surface value. When they were reminded that coal had been found long ago under the same
limestone at Ferry Hill, and bade to look at the Hetton
Colliery near them and judge for themselves, they replied,
that there was no reason to expect any coal about Haswell,
at any rate none better than had been found at Ferry Hill;
that the most eminent coal viewers of Newcastle had considered the matter, and decided that *if coal did exist* under
these estates it must be *of bad quality,* unfit to send to
London, and unworthy of the cost of sinking a deep pit.
To all this Mr. Smith could only oppose general but sound
geological truth; and though it convinced Colonel Braddyll
of the great value of his property, and had its influence
upon other landowners near, the ancient Newcastle prejudice of "no coal under the magnesian limestone," like the
Somersetshire and Lancashire prejudice of "no coal under
red earth," was not easily abandoned by the "practical"
men who had encouraged the delusion. Colonel Braddyll,
however, once convinced by sound arguments of the just-
ness of Mr. Smith's views, never relaxed in his efforts to

give them effect; and after some years of difficulty succeeded in establishing the magnificent works called the
South Hetton Colliery, which rival the mightiest establish-
ments of the Lambtons, Vanes and Russells.

This great enterprise might never have been undertaken,
this most profitable establishment might never have been
formed, but for the resolute honesty which enforced Mr.
Smith to press upon the owner of the estate the importance
of believing a scientific truth in opposition to a rooted error
of opinion in the district, and to the sentiments of persons
who might think themselves privileged to advise; yet the
execution of the grand scheme, for which Mr. Smith had
strenuously contended, was entrusted to others. The satisfaction of having done his duty under difficult circumstances,
and shown a convincing example of the value of his geological discoveries when applied to practical problems, was,
however, a great reward. He had, indeed, indulged the
expectation of benefiting in another manner by the success
of the undertaking which he so fearlessly recommended.

Kirkby Lonsdale was the next centre of Mr. Smith's operations, and here he had the great happiness (such he
always deemed it) of meeting (for the first time) Professor
Sedgwick, who had crossed over the hills from Teesdale.
The stone-masons in the romantic little town were keen observers of Mr. Smith's singular habits of handling the stones
and trying their hardness against his teeth, and guessing by
the hammer, which was not concealed by the Professor, that
he was "one of the same trade," immediately pointed the way
to his brother geologist. They walked together a short distance on the road toward Kendal, where (at Kestwick) Mr.
Smith had recently found organic remains (Orbicula, Or-
thoceras, &c.) in the quarries of what was then called
"grauwacke," and the Professor then continued his eques-
trian journey to Low Furness*.

The situation of Kirkby Lonsdale is most agreeable and

* A few days previously the writer had the good fortune to meet the
same celebrated Professor in a characteristic locality, within sight of the
High Force in Teesdale, and under characteristic circumstances; for he

interesting for the geologist; within a moderate walk a great series of rocks may be seen between the new red conglomerate and the upper slates. The strata are manifested in a variety of dislocated positions—vertical, highly inclined, horizontal, and contorted; fossils are abundant in the limestone rocks, and the botanist may contemplate the distribution of plants from the level of the sea to the summit of Ingleborough or Hougill Fells, on soils belonging to five distinct geological formations. Dry paths in several directions conduct to points which command the noblest views of rock, river, vale and mountain scenery, many of them surpassing in grandeur the well-known scene which expands beyond the terrace of the churchyard.

In this sweet retirement Mr. Smith long and willingly lingered, feeding all the best qualities of his mind by calm meditations, not unmixed with the poetic impressions which seem perpetually to haunt the romantic banks of the Lune. Frequent walks made all the neighbourhood familiar to us for a circle of fifteen miles, and gave the opportunity of completing the maps of Westmoreland and Lancashire, and making a great improvement on the map of Yorkshire, by introducing the slate rocks (Silurian) which in so remarkable a manner support the cone of Ingleborough, and the broader summit of Graygarth.

Mr. Smith took great pains to understand fully the coal-basin of Burton in Lonsdale, and the mixed series of grits, shales, coal and limestone, which overlies the great limestone of Farlton Knot and Hutton Roof. Fossils were collected in great numbers from the upper beds of the slate series at Beckfoot *in situ*, and gathered from bouldered masses of the same rocks on Kirkby Moor, and the limestone lands near Hutton Roof. Innumerable rambles led us up every glen and across every hill, now sketching waterfalls, anon tracing the boundaries of rocks, or marking the direction of " diluvial" detritus. Once or twice the radius of our orbit was

was riding, as usual, with saddle-bags for his specimens, and a miner's boy *en croupe*, who had promised to guide him to *a place where the limestone was turned into lump-sugar.*

lengthened by professional occupation, to include Durham, returning through Hellgill, and over Wild Boar Fell; to Garstang and the Vale of the Hodder; to Coniston and Low Furness; but we always returned with augmented delight to the quiet anchorage by the Lune. Here several of the latest county maps were finished, and a great variety of data collected, which have been found serviceable in after years. Many papers on geology and the economical applications of it, which Mr. Smith always desired to recommend and bring into daily use, were written at Kirkby Lonsdale.

A visit to Hesket Newmarket in the winter of 1822–23, interrupted this thoughtful, perhaps indolent repose; and for two or three months we were incessantly occupied by investigations of the lead and copper mines actually working, or of which ancient traces remained, in High Pike, Carrock, and the Caldbeck Fells. These mines, situated in a low part of the series of Cumbrian slates, and in the vicinity of the sienites of Carrock Fell, were established on veins in many respects similar to those of Cornwall, the matrix being generally quartz, the copper ores and lead ores sulphuretted in the deep parts of the mine, but oxidized and combined with mineral acids near the surface, where in places the back of the veins resembled the "gossan" so often described in Cornwall. Arseniuretted iron pyrites, wolfram, apatite, chrysocolla, arseniates, phosphates, carbonates of lead, and various other minerals, render this mining field one of the most curious in the North of England, and the general resemblance of it to the greater and richer districts of Cornwall is increased by the character of some felspatho-porphyritic (elvan) dykes which traverse the hills.

A pretty series of solitary crystals of sulphuret of iron, remarkably perfect, was collected from the interior of amorphous masses of the double sulphuret of iron and copper, which constituted the "ore" of one of these mines, and this reminded us of a curious fact observed at Holm Bank, near Conishead Priory, where, in the cupriferous joints of the limestone, crystals having the characteristic forms of iron pyrites, were found imbedded in and adjoining to carbonates of copper, themselves being brown oxide of iron. In the smelting house at Driggeth (under Carrock Fell) the process of reducing the argento-cupriferous lead ore had been so imperfectly performed, that a large weight (some tons) of mixed and sulphuretted metals (called " speltre" by the workmen) was taken off the top of the melted lead, by its adhering to the cold iron bar which was employed to stir the fused metal *. These mines of Caldbeck Fells had formerly been worked by Germans, and many of their old levels had been injudiciously abandoned by their successors.

In the examinations which Mr. Smith made of this interesting district, the writer was closely associated, and was stimulated by the extraordinary variety of the minerals in the veins and in the sienitic and porphyritic rocks, to investigate their crystalline structures and chemical composition theoretically and experimentally. On returning to Kirkby Lonsdale (Jan. 1823) this subject became a favourite and engrossing object of attention, and a small smelting-furnace (if such a title can be permitted to an apparatus consisting of only a few crucibles, and a water-bellows worked by hand) was kept in frequent use in a garden, surrounded by walls which were covered by fossils, rocks and minerals. Thus pleasantly passed days, months, almost years, in seclusion so entire, that only by indirect means and long after the date of the application, did Mr. Smith hear of an urgent official demand for his services as a mineral surveyor in Russia. He would probably have accepted the proposal had it reached him in time, but another and probably fitter destiny was reserved for him by Providence. Among the residents of Kirkby Lonsdale interested in the unusual processes of the wayward geologist, may be named Mr. Edward Wilson, who benevolently wished to benefit his native town by a better

* The galena was rather rich in silver, but somewhat difficult to free from copper, a very minute dose of which was found to harden the lead remarkably. It lay in the quartzose vein in a series of oblique (not vertical) discontinuous masses, as in Cornwall.

supply of water, obtained from springs in the neighbouring limestone rock. In this matter Mr. Smith assisted with his advice.

On one occasion, early in 1824, a relative of Mr. Wilson, Dr. Matthew Allen of York, visited Kirkby Lonsdale, was introduced to Mr. Smith, and on his return home mentioned the circumstance to some members of the Yorkshire Philosophical Society, then lately formed by gentlemen who had fortunately obtained a valuable series of organic remains from Kirkdale Cave. The consequence was an application from the President of the Society (the Rev. Wm. Vernon Harcourt) to Mr. Smith, to deliver a course of lectures on geology in York, and Mr. Smith, who though he had never *lectured*, had for half his life been *talking* on geology, immediately accepted the proposal. New maps were coloured, new sections drawn, and even the distant cabinet of Mr. Richardson at Farley was laid under contribution to supply illustrations for these discourses. On his arrival in York, Mr. Smith was warmly greeted by the zealous President and first Secretaries (Dr. Goldie and Mr. Copsie) of the Society; the lecture-room, fitted up for the occasion, was crowded; and the course was successfully conducted. The nature and order of the subjects discussed in these eight lectures may be in some degree judged of from the subjoined syllabus; but it must not be supposed that the formula was rigidly adhered to :—

" Syllabus of Lectures on Geology, by William Smith,
Mineral Surveyor.

" Lecture I.

" Introductory.—Geology a Science of great extent and universal interest; not a science of hard names, but of beautifully according facts.—The great facilities for acquiring it afforded in our own country.—Inducements to the study of Geology.—A view of its application to the purposes of human convenience.—Explanation of first principles of the Science, as developed in the original investigation around Bath.—Enumeration and description of the

Strata, illustrated by Sections and Maps; and the distinction explained between the alluvial and stratified matter.

" Lecture II.

" Connection of British Strata with those of adjacent countries.

" Exterior form of Land, as influenced by the terminations of the Strata in the Sea.—Bays, Estuaries, Capes, Cliffs, Low Shores, mutability of the Coast.

" Interior of the Island.—Principal ranges of high and low ground, contours of Hills, and characteristic forms of Mountains and Valleys, illustrated by drawings.

" Lecture III.

" Principles of natural Drainage.—1. Summits of Drainage; 2. Theory of Springs; 3. Courses of Streams, illustrated by reference to particular streams; 4. Characters of the Streams and Valleys in the respective Strata.—Lakes.—Interior Marshes.

" Sites of Population.—Origin of Towns, Villages, Castles, Abbeys, Seats, and Parks, as influenced by advantages arising from properties of the Strata.—Picturesque Scenery.—Varieties of Soil, and their agricultural appropriations.

" Locality of Plants and Free Animals, in the sea and on the land.

" Lecture IV.

" Strata identified by their imbedded Organized Fossils, and other remarkable bodies.—Use and application of the organized Fossils, in distinguishing one Stratum from another, exemplified by specimens and drawings, and reference to localities.—State of the preservation of Fossils, their relative antiquity, and resemblance to recent Shells, Corallines, &c.

" Lecture V.

" Variations from the general regularity of Strata.—Isolated masses, attenuation and discontinuity of Strata; remarkable elevations and depressions; unconformity of course and declination, and apparent dislocations of Strata.

—Dykes and Mineral Veins. Illustrated by sections and drawings.

" Lecture VI.

" Geology of Yorkshire.—Exemplifications of the great principles of the Science, around the coast and through the interior of the county.—Enumeration and description of the Yorkshire Strata, with their included organized Fossils and other peculiarities.—Instances of elevations and depressions, and other unconformities of the Strata, illustrated by Maps, Sections, and Drawings.

" Lecture VII.

" On the practical advantages of Geology, and its beneficial application to Agriculture, Mining, Coalworking, and Commerce in Yorkshire.—Valuable products of Yorkshire: Building Stone, Paving Stone, Slates, Marbles, Bricks, Pottery, Casting Sand, Grindstones, &c.—Sites of Population and Manufacture.—Discovery and working of Coal and Minerals.

" Lecture VIII.

" Review of the subject, with reflections on the formation of the Strata.—Proved to have been deposited from water liable to occasional agitation.—Change of climate considered. —Consolidation of Strata.—Effects of the Deluge, in detaching fragments from the rocks, rounding them by attrition in water, and scattering them over other Strata at great distances.—Bones of Land Animals found in this gravel, and in Limestone Caves.—Account of the Caves of Kirkdale, of Oreston, &c.—Antediluvian inhabitants of the earth.—Conclusion."

A certain abstractedness of mind, generated by long and solitary meditation, a habit of following out his own thoughts into new trains of research, even while engaged in explaining the simplest facts, continually broke the symmetry of Mr. Smith's lectures. Slight matters, things curious in themselves but not clearly or commonly associated with the general purpose of the lecture, swelled into excrescences, and stopped the growth of parts which were more important in themselves, or necessary to connect the observations into an intelligible and satisfactory system. But there was a charm thrown over these discourses by the novelty and appropriateness of the diagrams and modellings which exemplified the arrangement of rocks, the total absence of all technical trifling from the explanations, and the simplicity and earnestness of the man.

Mr. Smith made an excursion from York to the celebrated cave at Kirkdale, in company with the late Mr. Wm. Salmond, one of its most diligent explorers, and in the summer (1824) delivered a course of lectures on geology in the Town Hall at Scarborough, assisted by the writer. Always charmed with the bold and varied line of rocky coast, and interested by the geological peculiarities of the north-eastern part of Yorkshire, he gladly seized this occasion of renewing his residence at Scarborough, and his intercourse with several friends. At this time the well-known naturalist, Mr. Wm. Bean, who had long possessed a magnificent collection of the Testacea and other recent productions of the Scarborough coast, had begun to gather that large and exquisite series of organic remains which now enriches his beautiful museum. Mr. John Williamson was in full activity, and had already laid the foundation of the collection afterwards transferred to the museum of the Scarborough Philosophical Society. The rich hoard of Speeton fossils was almost untouched,— almost unexamined since the days of Lister,—the plants of Gristhorpe were unknown, but a spirit of geological research began to spread among the residents and visitors of Scarborough that promised the happiest fruits.

In this pleasant excitement no man had a truer delight than Mr. Smith. He loved to wander beneath the cliffs, noting the minutest variations in the stratification, detecting the slightest marks of dislocation, watching the peculiarities of the sea's action on materials of unlike qualities, and inferring the causes which had anciently modified the outline of the land, and covered the low cliffs of the oolitic series with fragments of the lias from Whitby, of the coal and limestone from Teesdale or Swaledale, and of the granite and sienite from Shap Fells and Carrock Pike. In numerous papers dedicated to the local geology of Scarborough his reflections on these subjects are recorded; and his exertions in examining one curious case of dislocation on the north side of the Castle Hill, brought on a rheumatic, or rather a paralytic affection of the muscles of the lower extremities, which bound him a prisoner in bed in the early part of 1825.

Previous to this accident he had taken part in a course of lectures to the Literary and Philosophical Society of Hull; after it had occurred, and before its effects were removed, while yet he was incapable of walking, and was actually lifted into the carriage which took him away, he accepted and executed a similar engagement, proposed by the Literary and Philosophical Society of Sheffield. It was a singular spectacle, to witness the delivery of lectures which required continual reference to large maps and numerous diagrams, by a man who could not stand, but was forced to read his address from a chair, to an audience of several hundred persons in a room not very well adapted for the voice. But it was far more extraordinary to witness during all the severity of the disorder, the unpretending patience and fortitude of the sufferer, who, had he then permitted his mind to dwell too curiously on the state of his health and the state of his finances, might have added the bitter foretaste of want and privation to the actual difficulty of the moment. Such reflections and such anticipations might sadden the hearts of those who surrounded him, but Mr. Smith would have thought it unworthy of his resolved mind and firm trust in Providence, to have abated one jot of his accustomed cheerfulness, shortened one of the innumerable playful stories which were always springing to his lips from the rich treasure-house of his memory, or turned his meditations from his favourite subjects.

At Sheffield, while slowly recovering the use of his limbs, he busied himself in arranging a body of information which he had gathered concerning the neighbouring coal districts;

and on removing soon afterwards to his old quarters at Don-caster, he worked much on the large Old Survey of York-shire, thinking to complete the colouring of it. By degrees he recovered entirely from his painful disorder, and from this year (1825) to 1839, nothing of the kind ever affected him again.

One of many agreeable incidents which diversified the repose of Mr. Smith's life at Scarborough, was a visit of Mr. Murchison to this part of the coast in 1826. The methods followed by this indefatigable geologist in his ex-tensive explorations, were such as to harmonize in several respects with the strictness of survey and exactitude of ob-servation which were always held by Mr. Smith to be essen-tial to a good field geologist.

In a trip by boat from Scarborough to Whitby, Mr. Smith was happy to point out to Mr. Murchison the principal results of his comparison of the oolitic and lias strata of Yorkshire with those of the South of England, and soon after received with great satisfaction the memoir on the oolitic coalfield of Brora, which illustrates in a remarkable manner the importance of such investigations.

Thus pleasantly passed two years in contemplating some of the most interesting scenes in Britain, and in meditating on the geological causes to which their singular beauty and variety could be referred. Some professional engagements at Scarborough and in other parts of Yorkshire occasionally drew Mr. Smith's attention, and he was especially interested in arranging a plan for a better supply of water to Scarbo-rough. The deficiency of water in the summer season, when Scarborough is full of visitors, was extreme; the structure of the promontory on which the town and castle stand is such as to yield only a very inconsiderable quantity of water on the spot; and the distant springs, which, by reason of their higher level, might be available for the supply of the town, are in general beyond the control of the corporation.

One small spring on Falsgrave Moor was at the disposal of the engineer, and it was in the economizing of this scanty rill that Mr. Smith exercised his ingenuity. He applied this

ingenuity at each end of the watercourse. In the hill he
excavated a subterranean reservoir, which might preserve
from surface waste the water at its origin, and in the town
he constructed a large closed reservoir, to meet the irregular demand of a varying population. These plans were successful, the little spring became a great blessing; the town
is benefited; but it has grown in size, and now requires a
further effort to obtain further supplies from another source.

Among the many eminent persons who at different periods
of Mr. Smith's life took a lively interest in his welfare, it is
the pleasing duty of his biographer to mark with grateful
distinction one whose friendly regard he gained about this
period, and retained during the remainder of his life, Sir
John V. B. Johnstone, Bart., of Hackness. On succeeding
to his estates, this enlightened gentleman was desirous of
converting to practical effect on his farms some of the geological and botanical truths which he knew to have been
established in the museum and the laboratory ; he found in
Mr. Smith the union of practical and theoretical knowledge
which was necessary for his object, and a desire to exemplify
that knowledge in agricultural improvements which exactly
coincided with his own wishes. From 1828 to 1834 Mr.
Smith acted as his land-steward, resided at Hackness, occupied himself in the usual concerns of a large landed estate,
and thus passed (in the judgment of the writer) six of the
calmest and happiest years of his declining life. The worthy
proprietor of Hackness had hoped that the retirement which
seemed so well suited to Mr. Smith's age and taste, would
have been memorable for the production of the results of a
life of scientific toil, and spared neither friendly entreaty,
nor pecuniary aid, nor personal exertion, to bring this fa-
vorite design to effect. Mr. Smith meditated and wrote,
but did not arrange his papers, and excepting a beautiful
geological map of the Hackness estate, executed in great
detail and with extreme exactitude, nothing of importance
came from his hands to the public. He was now advanced
to the age of sixty years: forty of these years had been
passed in almost solitary research and intense meditation:

I

the publications which he had accomplished had been produced under circumstances of great difficulty and discou-
ragement, and had yielded him no adequate return. Perhaps, under these conditions, it was too much to expect of a
man, whose happiness lay in the solitary world of his own
thoughts, that he should now correct the one error of his
life, and forgo, at the call of duty, the habit of procrastina-
tion, which had resisted the ambition of scientific fame, and
the pressure of actual distress.

But these years were fruitful of events interesting to
the friends of William Smith. In Feb. 1831, the Council of
the Geological Society of London honoured him by awarding to him the first Wollaston medal; and the terms with
which the gift was accompanied render this act on the
part of the Society and the President extremely memorable.
Dr. Wollaston's services to physical science were well known
and duly honoured in his life-time; geology has felt, and
will long feel the benefit of his dying bequest. He invested
one thousand pounds in the three per cent. Reduced Bank
Annuities, in the joint names of himself and the Geological
Society, and directed that after his decease, " the Society
should apply the dividends in promoting researches concerning the mineral structure of the earth, or in rewarding those
by whom such researches should hereafter be made; or in
such manner as should appear to the Council of the said
Society for the time being, conducive to the interests of the
Society in particular, or the science of geology in general."
He afterwards enjoined the Society " not to hoard the dividends parsimoniously, but to expend them liberally, and, as
far as might be, annually, in furthering the objects of the
trust." The first year's income from this fund was appro-
priated to the acquisition of a die for a medal bearing the
head of Dr. Wollaston, and this having been undertaken by
Mr. Wyon, the Society was prepared in 1831 to fulfil the
trust with which they were charged. The Council accordingly passed *unanimously* the following resolutions, Jan. 11,
1831 :—

1. " That a medal of fine gold, bearing the impress of the

head of Dr. Wollaston, and not exceeding the value of ten guineas, be procured with the least possible delay.

2. " That the first Wollaston medal be given to Mr. William Smith, in consideration of his being a great original discoverer in English geology; and especially for his having been the first, in this country, to discover and to teach the identification of strata, and to determine their succession by means of their imbedded fossils *."

The announcement of this award was made by a congenial spirit. The chair of the Geological Society was then filled by one of its most honoured members, an original thinker and faithful observer, well qualified to appreciate the originality of Mr. Smith's discoveries, and well acquainted by actual research with their extent and their value. In his address on this occasion, Professor Sedgwick, speaking in the name of the Geological Society, sketched a brief but satisfactory history of Mr. Smith's career, demonstrated the entire justice of the award of the Council of the Geological Society, and added his personal testimony in favour of Mr. Smith's claims in terms of no ordinary value.

" I for one can speak with gratitude of the practical lessons I have received from Mr. Smith: it was by tracking his footsteps, with his maps in my hand, through Wiltshire and the neighbouring counties, where he had trodden nearly thirty years before, that I first learned the subdivisions of our oolitic series, and apprehended the meaning of those arbitrary and somewhat uncouth terms, which we derive from him as our master, which have long become engrafted into the conventional language of English geologists, and through their influence have been, in part, also adopted by the naturalists of the Continent.

"After such a statement, gentlemen, I have a right to speak boldly, and to demand your approbation of the Council's award. I could almost dare to wish that stern lover of truth, to whose bounty we owe the ' Donation Fund,' that

* In the medal presented to Mr. Smith is a very slight imperfection, arising from a minute want of finish in the die, which has been since removed. That imperfection may be valued as authenticating the "first" medal.

dark eye, before the glance of which all false pretensions
withered, were once more amongst us. And if it be denied
us to hope that a spirit like that of Wollaston should often
be embodied on the earth, I would appeal to those intelligent men who form the strength and ornament of this Society, whether there was any place for doubt or hesitation?
whether we were not compelled, by every motive which the
judgment can approve and the heart can sanction, to perform this act of filial duty, before we thought of the claims
of any other man, and to place our first honour on the brow
of the father of English geology?

" If, in the pride of our present strength, we were disposed
to forget our origin, our very speech would bewray us: for
we use the language which he taught us in the infancy of
our science. If we, by our united efforts, are chiseling the
ornaments and slowly raising up the pinnacles of one of the
temples of nature, it was he who gave the plan, and laid the
foundations, and erected a portion of the solid walls by the
unassisted labour of his hands.

" The men who have led the way in useful discoveries
have ever held the first place of honour in the estimation of
all who in aftertimes have understood their works or trodden
in their steps. It is upon this abiding principle that we have
acted; and in awarding our first prize to Mr. Smith, we
believe that we have done honour to our own body, and are
sanctioned by the highest feelings which bind societies
together.

" I think it a high privilege to fill this chair on an occasion
when we are met, not coldly to deliberate on the balance of
conflicting claims, in which, after all, we might go wrong,
and give the prize to one man by injustice to another; but
to perform a sacred duty, where there is no room for doubt
or error, and to perform an act of public gratitude, in which
the judgment and the feelings are united*."

On this occasion Mr. Smith presented to the Society the
original Table of Stratification drawn up in 1799, referred

* Address to the Geological Society, at the Anniversary Meeting, Feb. 18,
1831, by the President, the Rev. Adam Sedgwick, M.A., F.R.S., &c.

to in this work (page 30), and a circle-map of the vicinity of Bath, which had been geologically coloured about the same period.

The British Association, founded at York in 1831, held its second meeting at Oxford in June 1832; and on this occasion the Wollaston Medal, awarded in the previous year, was put in Mr. Smith's possession; and he was further gratified by the announcement that a pension of £100, solicited by the united voice of English geologists, had been assigned him by the Government of His Majesty William the Fourth. Thus a second time in the progress of this memoir have we found the British Government yielding to the honest influence of lovers of science, and the friendly interference of men who adorned their official stations, and marking with favour the cultivation of a branch of the knowledge of nature, which is far from being, even now, appreciated by the country at large. Those who freely censure constitutional governments for the tardiness and feebleness of their patronage of science, should number among the causes of this seeming indifference, the want, among the friends and cultivators of knowledge, of the union of effort and the singleness of object which have been found effective in guarding the interests of other portions of the community. Such union it was one of the great purposes of the British Association to accomplish and perpetuate; and the resolute support which it has received from the cultivators of natural knowledge in the British Isles, and the attention which has been on several occasions vouchsafed to it by the Government, appear to prove that there is no particular difficulty in harmonizing the efforts of scientific men, nor any unreasonable obstacle to a just consideration of their labours in the channels of official life.

A further gratifying proof that the attention of the Government was awakened to the interests of science was given to the members of the British Association at their meeting in Cambridge in 1833, when Dr. Dalton's name, long ennobled by the admiration of Europe, was placed on the Pension List for £300.

To Mr. Smith the periodical return of these meetings was like the revival of spring to the vegetable world. Except at their summons, he rarely quitted his hermitage at Hackness, and for many years his presence in the Sections of the Association was hailed with delight by those who, occupying the highest place in public opinion, were proud to call him the "Father of Geology," though to some of the views which they advocated he was disposed to show anything but parental affection. But he rarely took part in discussions at these meetings, and indeed seldom spoke, except to mention some striking fact, because his mind had been little trained to intellectual gymnastics, and an infirmity of age—the only failing which age brought upon him—deafness, began to rob him of half the wit and eloquence which he most admired and most longed to hear from the lips of Professor Sedgwick.

In 1834, on occasion of attending the Edinburgh meeting of the Association, he was a visitor to Sir Charles Monteith, and made himself well acquainted with the singular features of the geology of Closeburn and the Valley of the Nith. Among other specimens which he brought from Nithsdale is a remarkable mass of Silurian sandstone, very rich in corals and shells, such as are well known in Shropshire and South Wales. The locality is marked on it, but not in the precise manner which was Mr. Smith's wont in earlier days. He also explored some parts of the Highland district of Scotland, and made other excursions from Hackness, which in this year he quitted to resume his residence at Scarborough.

The following notice on this subject must not be omitted:

"Scarborough, June 28, 1839.

"After living at Hackness near six years I grew weary of nothing but farming concerns, and told my good friend Sir John Johnstone I wished to leave it, and that as the last five or six years of a man's life were seldom good for much, I wished to have them to myself (provided I lived so long), to complete and arrange my papers without the interruption of any business, to which he readily acceded, and kindly allowed me twenty pounds a year for occasional advice and visits."

The prediction, if it may be so called, which this sentence contains was verified; Mr. Smith did live five years after quitting Hackness, and he employed himself with industry on his papers, less however to arrange than to augment their already voluminous mass. At Hackness, however, he had not been so wholly engrossed by "farming concerns" as he perhaps imagined, for a great number of papers on various subjects, which imply the exercise of a mind much at ease in respect of occupation, and capable of dwelling on such reflections as retirement and leisure produce, are dated "Hackness."

From this time, unfettered by any but temporary engagements, which left him the free disposal of abundant leisure, Mr. Smith remained principally at Scarborough, and devoted his mind to a review of the circumstances of his life and the arrangement of his observations and opinions. In a locality which he loved, a retirement which he chose, and under the agreeable stimulus of assured respect and widening reputation, his hours passed almost unheeded by. Buried in the seclusion of his study full of his own maps and manuscripts, or wandering beneath the cliffs whose geological structure he was the first to comprehend, his mind was strung anew, and it might have been expected that the thousands of facts which his memory retained should have been methodized into laws of phænomena, and the characteristic and unusual inferences which he expressed, on the theory and economical applications of geology, have acquired a systematic and permanent form. Such were his own expectations, and it was owing rather to a discursive habit of mind, an excessive activity of observation which seized upon and gave temporary importance to every novelty, than to any want of industry or indecision of judgment, that these apparently well-founded hopes were not fulfilled. Papers which required arrangement and condensation were augmented, and connected with new trains of research ; and, as if the ordinary forms of expression were inadequate to match the vividness of his ideas connected with the structure of the earth, his thoughts on stratification and organic remains were often clothed in humourous if not always harmonious verse. Those to whom this play of fancy was revealed easily abandoned the belief that Mr. Smith's geological principles would ever become fully known through his own exertions; and as no inducement could persuade him to intrust to others the task which he felt it was his duty to accomplish, the case became from year to year more desperate, as the observations lost more and more of their freshness, and the generalizations of their novelty. Mr. Smith mixed too little with younger geologists to make the discovery, that on the road which he opened were racers swifter than himself; and few of his friends could undertake the painful task of proving to him the unpleasant truth, that the piles of manuscript, in which he fondly hoped that both fame and fortune were secured, were rapidly losing their value as contributions to science and articles of trade.

The monotony of his life was often relieved by visits of a few weeks' duration to London, where he enjoyed to some extent intercourse with the active members of the Geological Society, and attended with interest a few scientific lectures. Increasing deafness, however, deprived him of a full share of these enjoyments, and his life in London, whither he always carried large packages of manuscript, was little different in its objects and results from that which has been described at Scarborough.

Mr. Smith was received at the Dublin meeting of the British Association in 1835 with marked kindness by Dr. Lloyd, the good Provost, and many of the eminent Fellows of Trinity College ; and this noble establishment conferred on him an honour which was entirely unexpected, the degree of LL.D. Dr. Smith, as we must now call him, was, in his sixty-seventh year, sufficiently alive to feelings more common in youth to be highly gratified with his new title of honour, and his friends, both amused and pleased with this natural and grateful exultation, strove by every means to augment his enjoyment. The Rev. J. Maclean, Dean of the College, showed him the kindest and most persevering attention, conducted him to many interesting scenes, and made his journey through the North of Ireland one of continual delight. Captain Portlock, who then directed the operations of the Geological Survey in Ireland, and was stationed at Belfast, took particular pleasure in aiding the inquiries of the veteran geologist, whose mind, still active and benevolent, was turned at this time to questions which no one could so well handle, questions of drainage and agricultural improvement, of much importance to Ireland. Perhaps some of the speculations in which he indulged might have been turned to valuable practice but for the death of one of the kindest of all the friends of his later life, Mr. Maclean, an event much and generally deplored.

Dr. Smith was called to attend the meeting of the Association in 1836 by a special invitation from the inhabitants of Bristol, who defrayed the cost of his journey. On this occasion he renewed his old friendship with Mrs. Richardson, and Mr. Wm. Pyle Taunton and his lady, the daughter of the Rev. Joseph Townsend, and was enabled to repeat his examination of points which he had not seen for thirty years. At this time he occupied some days in a rather minute and careful survey of part of the coal district of Kingswood, and coloured the result of his observations on the sheets of the Ordnance Survey.

Amongst a few professional engagements which occupied Dr. Smith's attention in 1837 and 1838, it may be sufficient to mention one, the most remarkable of all, in which, by direction of the Government, he was associated with Mr. Barry and Mr. (now Sir Henry) De la Beche. Before proceeding to the execution of Mr. Barry's magnificent design for the new houses of Parliament, and making choice of the stone to be employed in the building, it appeared desirable that all the principal quarries of the kingdom should be examined, and the qualities of the stone most celebrated in decorative architecture studied in edifices, examined in the laboratory, and subjected to mechanical experiment. From such an investigation might be expected not only a good selection, on sure grounds, of a proper stone for the Parliament-houses, but general rules founded in practice and theory, and applicable to a variety of architectural and engineering operations. These expectations were fulfilled.

Leaving Newcastle in 1838 (at the close of the meeting of the British Association), the party traversed rapidly a great part of England and Wales in the months of August, September and October, and collected a mass of information regarding the situation of quarries, the cost and "workability" of the stones most in reputation among builders, and the evidence of their durability in edifices of authenticated antiquity. Specimens of these various sorts of stone were subjected to chemical analysis, and to experiments on specific gravity, power of cohesion, and hygrometric properties, by Professors Daniell and Wheatstone, and the result of the whole inquiry was published in a detailed Report to the Commissioners of Her Majesty's Woods, Forests, Land Revenues, Works and Buildings, dated 16th March 1839. This valuable document consists principally of tabulated observations and experiments relating to above 100 quarries in Scotland, England and Wales, and is prefaced by remarks on the nature and result of the inquiry, which deserve the attention of scientific and practical men. The stone selected was the firm, yellow, granular magnesian limestone of Bolsover Moor in Derbyshire ; and examples of this and many other sorts of stone, of various degrees of excellence and suited to various uses, may be seen in the Museum of Economic Geology, Craig's-court, Charing-cross, established, under the direction of Sir H. T. De la Beche, by the Commissioners of Woods and Forests.

This was an investigation to which Dr. Smith willingly gave all the earnest attention which it merited, and his previous knowledge of nearly all the building-stones and quarries in the kingdom was found highly beneficial to the Commission. Five days after the signing of the Report (in London) he attained his seventieth year, in excellent health and spirits, which appeared to promise years of thought, if not of activity. Except a few days employed in re-examining with Mr. Barry the quarries near Worksop, which had been selected by the Commission, he remained principally in the vicinity of Scarborough (" walking over farms at Silpho and Broxa, and other parts of the Hackness estate") till July. In the early days of this month he was occupied in arrangements intended to secure a quiet possession of his pleasant house at Scarborough for some years more, and on the fifth he left it, never to return. Perhaps it was incautious in a man of his age and habits to venture an outside passage by the mail to London, but there is no evidence that he suffered by it. On the 10th he attended the English Agricultural Society in Cavendish-square; on the 11th he was engaged with Mr. Philip Pusey, the great mover of that association; the 16th saw him in Oxford, where he dined at Queen's College with 2500 persons; and the 18th brought him to the hospitality of Nuneham. Here, assisted in his desire to examine parts of the neighbouring country by the friendly regard which for twenty years had been manifested towards him by the Archbishop of York and many members of his family, Dr. Smith passed several pleasant days, visiting Watlington, Garsington, Milton, Marcham and Stanford, with an enthusiastic anxiety to complete in age the investigations among the oolites and chalk hills which he had commenced in youth. Afterwards he revisited for a few days his native place and rural friends, and returned to London on the 9th of August.

The meeting of the British Association was this year (1839) appointed to be held at Birmingham on the 26th of August, and Dr. Smith received from Mr. Joseph Hodgson (one of the secretaries of the meeting) a special and very cordial invitation to be present. He stopped on his journey to Birmingham at the house of his highly valued friends, Mr. George Baker, and Miss Baker of Northampton. Here the kindest welcome awaited him; and in addition to the pleasure of contemplating the beautiful series of Northamptonshire fossils which Mr. and Miss Baker had collected, he was gratified by several excursions into the neighbouring country, which had always been interesting to him since in earlier days he had opened the curious volume of Morton's 'Northamptonshire.' While thus tracing the boundaries of the minor divisions of the oolitic rocks which he had been the first to distinguish, a slight cold by which he was affected seemed, to the eyes of his friends, to deserve more attention than he bestowed on it; diarrhœa gradually supervened, and medical assistance became immediately advisable. Dr. Smith had for many years been successful in guarding his own usually robust health, and he was slow and reluctant to admit of advice better suited to the disorder which now attacked him, and which on a former occasion had so prostrated his strength that he recovered with difficulty under the treatment of his friend Mr. Dunn. He began to feel the attack serious, and to perceive the alarm in the faces of his friends, before Dr. Robertson of Northampton was called to his aid. The author, then earnestly expecting his revered relative at the meeting of the British Association in Birmingham, received information of his illness on the evening of the 26th of August, and in the morning of the 27th attended his bedside.

It was difficult to believe, that under that calm, thoughtful, and pleased expression of countenance, those animated descriptions of the country which he had visited a few days previously, those plans of further and strenuous exertion, which asked years of active life for completion, lurked pain and fatal disease. If there had been some trace of delirium, this had disappeared, and it seemed as if the remedies applied were producing beneficial effects, but this hope failed; the uncomplaining sufferer sunk continually in each succeeding hour, till his eyes lost their bright and kindly light, and the ever-varying features became fixed in serene and awful tranquillity (Aug. 28, 10 p.m.).

––––––––––

Dr. Smith was buried at Northampton, at the west end of the beautiful antique church of All Saints, in which, at the

suggestion of Dr. Buckland, a tablet will be placed to his
memory, by a subscription among geologists.

Several portraits of Dr. Smith exist, as the following account testifies :—

" Portraits.

" Scarborough, June 29, 1839.

" My portrait was first taken in 1805 by Solomon Wil-
liams, who painted the ' Trial of Sidney.' He was then about
painting the group of attendants at the Holkham sheepshearing. Some of the most eminent persons sat to him at
my house in Buckingham-street. The Duke of Norfolk
(then Bernard Howard), Lord Somerville, Sir Joseph Banks,
Thomas Crook, and some others, forming a group on the
left-hand side of the picture, were completed, and I, bare-
headed, was introduced behind Sir Joseph Banks ; but this
picture was never finished. He also painted another of me
in oil, somewhat larger, which I have, but that in the group
before mentioned was thought to be a better likeness.

" My friends at Scarborough wishing me to have my likeness well taken, gave me a letter of introduction to the celebrated self-taught artist of Lestingham in Yorkshire [Jack-
son], and when I went to London [1832] I sat to him, for
which he would take no remuneration; and on leaving it in
his hands to be engraved, I asked him what he thought of
it, when he said he believed it to be as good a likeness as
ever he drew in his life. He was, however, shortly after
taken ill, and died, so that this (only pencil stippling) was
sold at his sale, with a few other little things, for five
guineas.

" In the summer of 1838, a tall, well-grown, fine-looking
young gentleman from France, for a very short time became
an inmate at my lodgings, 6 Lancaster-place, Waterloo-
bridge, and had not been there more than three or four
days before he said he ' should like to take my portrait—it
would make a good picture—if I would permit.' I told him

I could not afford to pay for it. ' Oh,' says he, ' artists are never paid.' Consent being given, he said, ' Tomorrow me at 8, you at 10;' and accordingly in the morning at 10 I found in my room he had prepared the canvas, put on his painter's silk gown of all colours, adjusted the lights, placed me in one chair and himself in another, set to work, without any easel, and by 4 o'clock in the afternoon, with about half an hour's re-touching the next morning, he produced a fine oil painting.

" I never saw a man stick so closely to his task or handle his tools so dextrously. There was no time lost in idle conversation, for he could speak but few words of English, and I none of French.

" It was thus, by the skill and generosity of my much-esteemed young friend M. Fourau, I became possessed of a fine oil painting. He requested me to write on the back of it—*Portrait of Dr. William Smith, painted in London,* which I did in a strong hand."

The engraved portrait in this volume is copied from the picture of M. Fourau.

Dr. Smith's person was formed on large proportions, cor-responding to the sturdy strength of his intellect ; his health was generally good ; he enjoyed a sound mind in a sound body. When about forty years of age he suffered by ague, caught in the marshes of Laugharn, and at subsequent times by diarrhœa, probably through exposure to cold, which he always faced without a great-coat, or the defence of hand-shoes, as he contemptuously termed gloves. To slight attacks of lumbago he was rather subject, and from about the fiftieth to the sixtieth year he was troubled by a gravel complaint ; this was relieved by copious draughts of vegetable bitters (camomile tea), and by this treatment and extreme moderation in diet, followed by abstinence for a time from all fermented liquors, entirely cured, so that in the latter years of his life he was apparently and really free from illness. Descended from a healthy race, amongst whom tooth-ache was unknown and longevity was frequent—uncles and great-grandfathers having seen their 98th year—temperate in diet, moderate in exercise, and vigorous in mind, his whole life was a continuous stream of thought and action. He can hardly be said or supposed to have lost by neglect a single hour, though, by permitting his mind to wander from its appointed line of research into innumerable by-paths, he sometimes forgot or postponed, and therefore failed in his main object.

A remarkable quality in the character of Smith was firmness, which, according to the difficulties, disappointments and sorrows of his life, took the aspect of fortitude, patience, and resignation; but in the prosecution of scientific or professional labour, rose to cheerful courage and persevering resolution. The exercise of this high quality gained him in early life professional independence and scientific fame, and in later years preserved to him, amidst poverty and domestic affliction, a calm elastic mind, the envy of younger men.

The "Map of the Strata of England and Wales" is a monument of labour and judgment: if to this we add twenty-one geological maps of counties and many detailed sections, a great variety of reports in engineering and mineral surveying, and innumerable detached essays on geological and economical subjects, we shall grant to the author and writer the praise of unremitting industry. More abundant fruits of this industry would have been given to the world had he been more regularly trained, especially in literary pursuits; for thus the excessive aptitude of his mind for original and discursive research might have been directed in more methodical channels to more systematic and complete results.

The life of a professional man, and especially that of an engineer, is seldom favourable for the acquirement of the valuable habit of restricted and regular study: Smith's career was in all respects such as to render the exercise of this habit impracticable; his papers have been mostly written at short intervals of rest, while travelling, and are thus only fragments which the author alone could have arranged into an edifice—links of a broken chain which can never be re-united.

Voluminous as these papers are, they do not contain a vast variety of matters which had passed under Smith's observation. His memory retained whatever his eyes had seen; and it has often occurred to friends who listened to his precise and complete narrations of past events, to regret that no practised amanuensis was at hand to preserve much that ought not to have been lost. There are living geologists who may be warned by these remarks how to provide against the mischiefs of indolence or indisposition, and by the humble aid of swift writing, to save for future times those precious thoughts which else will exist only in the fading recollections of admiring friends.

Had Smith been asked what he thought the most prevailing quality of his mind, he would doubtless have replied, " a habit of observation."

" Scarborough, November 16, 1838.

" By these reminiscences I see how the habit of observation crept on me, gained a settlement in my mind, became a constant associate of my life, and started up in activity at the first thoughts of a journey; so that I generally went off well prepared with maps, and sometimes with contemplations on its objects, or on those on the road, reduced to writing, before it commenced.

" My mind was therefore like the canvas of a painter, well prepared for the first and best impressions."

This habit, founded on a natural quickness of the senses, was nourished in boyhood and cultivated through life. The eye was in him more than in other men the avenue of knowledge to the mind, and was educated with proportionate care. Often, when desirous of remembering a certain name, he would write it distinctly, saying, that if he forgot the sound he should remember the picture. This memory for form made him an easy and accurate sketcher, and, with a good taste for colour, gave him great enjoyment in paintings and sculptures. [See the portraits of the Duke of Bedford,

Rev. J. Townsend, and Dr. Anderson, in the former parts
of this volume.]

He exhibited on many occasions a considerable facility
in mechanical inventions. When the bore-hole in the pit
sunk at Batheaston had reached the lias, so much water
sprung up as to fill the pit, and overpower the engine.
What was to be done? Smith instantly met the case by
a very simple arrangement. He caused a long piece of wood
to be planed with eight sides tapering to a point; at the
large end he screwed on a heavy iron top, and to an eye in
this was fastened a rope of sufficient length and strength.
The machine (if it may be so called) was let down to the pit
bottom, and moved about in the water till the point of the
rod entered the bore-hole; it was then permitted to drop
into it; the iron head was unscrewed, lifted, and again permitted to fall as a hammer on the rod, which by three or
four blows became fixed in the hole. Thus the spring was
stopped, so that the engine, being set to work again, easily
emptied the pit.

While engaged at Norwich in preparing for the press his
work on Irrigation (1806), Smith's attention was drawn to
the processes of the typographer (Mr. R. Bacon), and he
conceived the idea of continuous printing, by causing an
inking apparatus to revolve against a revolving frame of
type; he drew diagrams and executed models of the requi-
site surfaces, the type-frame having four *plane* faces, and
the inking frame four faces *curved* so as to *meet them
exactly*. Mr. Bacon afterwards applied to Mr. B. Donkin
to arrange a printing apparatus on this general idea, but
under the hands of that eminent engineer it took quite a
different form.

The colouring of the great Map of the Strata was on a
new and peculiar plan, the terminal edges of the rocks being
deeply tinted, and the other parts of their visible surface
merely washed; the constructing and colouring of the sections were equally unlike what were adopted previously;
the very shelves on which his primitive collection was arranged bore the same impress of original and independent

K

thought. These were in wood what the sections were on paper, the boards being made all to *slope in one direction*, so as to imitate the prevalent inclination of the rocks; and their terminal edges rising to greater or less heights above the floor, according as the rocks which they represented formed hills or valleys on the earth's surface. This arrangement has been partially copied in the beautiful Rotunda at Scarborough, in the plan and execution of which on many points Smith's advice was followed.

A man whose life was passed in reducing to order the rugged aspect of the earth, might claim to be excused for any slight want of refinement in manner; but Dr. Smith's natural goodness of heart and variety of knowledge rendered his society agreeable to most persons, and highly attractive to those who valued him or the science he had unfolded. In his intercourse with such friends, the ' malus pudor' and the ' superbia quæsita meritis,' which seem to haunt men of genius, were driven away by an unrestrained flow of pleasant narration, or an earnest pleading for what was deemed true, and sacred because true. On such occasions, the resolution with which he held to an opinion once formed on one good, or on several plausible reasons, might deviate into prejudice, if the arguments brought to oppose him were not of such a nature as he deemed fitting for the case. He would yield always to a plain and clear statement of fact, but seldom to a demonstration involving or founded on the progress of collateral science. Geology was with him ' the science,' with its own classes of observations, arguments and conclusions. Such light as zoology, botany, chemistry, or mechanics could throw on the ancient phænomena of nature was admitted slowly and cautiously, unless independently sustained by direct observations on the strata; and as for their verification he would seldom trust others than himself, his views of geological theory were sometimes difficult and embarrassed.

A favourite topic of conversation with Dr. Smith was the history of his own geological researches, coupled with notices of his professional labours, and doubtless there are

many persons now living who have heard from his lips a
more full and circumstantial account of several events of his
life than it was necessary or indeed practicable to record in
these pages. They will also remember innumerable pleasant anecdotes and characteristic traits of many eminent
individuals with whom he had been associated. It is difficult to resist the temptation of reporting these amusing notices of events which happened half a century since, but a
very small selection must suffice.

One of his earliest and most esteemed friends, Mr. Thomas
Davies, steward for the Marquis of Bath at Longleat, who
was the means of introducing him to Mr. Richardson, was
remarkable for pithy sayings, in good Wiltshire idiom. During the palmy days of the Bath Agricultural Society (about
1800), he one day observed to a farmer at Longleat that he
had not seen him at the last agricultural meeting. " Why
noa, zur," replied Hodge; "I have been thinking, zur, these
agricultural meetings don't do much good." " I tell ye
what, my friend," said the steward, " they have done some
good if they have set you o' thinking, for that's what you
never did in your life before."

On occasion of a visit to Mr. Crawshay at Merthyr Tydvil
(1803), he had an opportunity of hearing the sentiments of
a brother ironmaster on the subject of teaching the art of
writing to poor children, then debated in the midst of the
furnaces. " No, no," exclaimed this really intelligent and
benevolent, if prejudiced man, " they'll all be hanged for
forgery; put 'em into a three-foot vein,"—a thickness of coal
esteemed to be suitable for " training up a child in the way
he should go !"

Being consulted by a landowner on the dry Mendip Hills,
as to the best means of procuring supplies of water for his
farm, Smith found that he had been anticipated in delivering an opinion on this very difficult subject by a miner, who
proposed to solve the problem by divination, that is to say,
by the divining or jowsing (chowsing?) rod. Unwilling
that his worthy employer should be at the mercy of this
superstition, he filled his pockets with some small stones not

K 2

commonly found on the Mendips, and proceeded to witness the trial of the " forked stick." Accordingly the miner exercised, in presence of the owner and the geologist, the "mystery" of the rod, and wherever the point of the twig turned downward, declared that water was to be found by digging. At these points Smith quietly dropped the stones, and when several places had been thus pitched upon asked the miner if he could rediscover the points indicated. Unaware of the stratagem, the man readily agreed to repeat the trials on the way home. In his progress he unluckily passed the spots where the stones lay and stopped at several other localities, to which the faithless rod directed attention; on which Smith remarked, that as the water had in so short a time changed its situation at all the points, it would be imprudent to spend money in following it.

On reviewing the course of William Smith's life and labours, we have been forcibly struck with the adaptation of his character to his position, and impressed with the difficulty of judging of a man's mental powers by his practical deeds,—of a man's real enjoyment of life by the merely external circumstances which surround him. Who that heard the hearty laugh of the " narrative old man" would suppose that he had lost " house and land," spent more than the hard earnings of his daily life, and endured of what is called affliction more than commonly falls to the lot of man? Who that knew *what* he had endured, could believe that, in the heaviest hour, he had never yielded to depression, never ceased to command his mind and turn its unfailing energy to the subjects he *willed* to contemplate? Any one, aware of the disadvantages from which he rose into renown, might suppose his great discoveries to have been the sudden result of accident, rather than the growth of exact observation and careful reflection. Yet in all these cases the first impressions would be utterly in error. Perhaps a greater error than all would be the supposition that, under more flourishing worldly circumstances, the labours of this remarkable man would have

been greatly more important. *He did not think so himself.*
They who argue thus are perhaps not right in their general
view of human nature, and certainly they are wrong in the
estimate they form of the character of Smith.   Hearts which
brave the attacks of adversity, yield to the solicitation of
prosperity.   There was in the luxurious musing which he
indulged somewhat of indolence, and it required often the
pressure of business and the lack of money to rouse him to
needful exertion.   Except by the peculiar profession which
he created for himself, the peculiar work which he set himself to do, the gathering of materials for his great Map,
could not have been accomplished by an individual.

It has been conjectured that the progress of English
geology would have been accelerated had the infant Geological Society of 1808 taken up Smith's principles, and
adopted his map as a basis of operations. *It is very im-
probable that he would have agreed to share with any man,
or any body of men, the labour and the honour of the work,*
on which he felt himself entitled to write, " Alone I did it."
Whether he was right in the opinion he formed of the value
and independence of his researches, and of his own power
to follow them out, is not now to be questioned; it may be,
however, worth while, before closing this record, to attempt
to mark with precision the place in the scale of geological
discovery which has been awarded to William Smith by the
unanimous vote of contemporary geologists *.

Very many *facts* are known by experience before the
laws which unite these facts into system are embodied into
science.   The stratification of many rocks; the alternation
of rocks of different nature; the peculiar positions of metallic veins; these and many other circumstances, which are
now known in the shape of general *laws of phænomena*, must
have been known in the mines and collieries and quarries
of Europe from an early period of the middle ages.   The
quality and distribution of soils; the characteristic features
of ranges of hills; the peculiarities of the origin and course

* To the geologists of England this subject is familiar through the admirable essay of Dr. Fitton, already alluded to.

of many rivers; these and other facts observable on the *surface of the earth*, which are now seen to depend on the peculiar qualities and positions of rocks below the surface, were formerly known as insulated facts, and, in some instances, may have suggested to such men as Packe, and Lister, and Evelyn, views of the earth's structure more conformable to modern philosophy than appear in their writings.

The naturalists of Italy, represented by Scilla, had begun to reason correctly on several points of the history of organic remains before our Woodward mixed truth and error in his 'Natural History of the Earth,' and threw into unprofitable discussions about the Deluge, the talent which had been awakened in England to contemplate facts which are the foundation of the highest geological laws.

The glimpses of general truth which those early writers obtained constitute a considerable body of knowledge, and are far more worthy of a place in the history of English geology than the sounding speculations of Burnet, and Whiston, and even Whitehurst, great as are the merits in other respects of that remarkable author.

That the earth is *stratified* in parts near the surface was generally known to the writers on geology in the sixteenth century, and is indeed one of the leading phænomena which their hypotheses were framed to explain. That the strata, considered as definitely extended masses, were arranged one upon another in a certain *settled order* or *series*, nowhere appears as a law, even for a limited district, previous to the remarkable paper by John Strachey in the Phil. Trans., which gives the succession of strata in the coal district of Somersetshire (1719 and 1725). For here (1719) we have the coal series covered, and covered unconformably, by the red marl, lias and oolite; and (in 1725) the limits of the coal-field, between Mendip Hills on the south, Cotswold on the north-east, and Marlborough Downs or Salisbury Plain. In a section he represents, in their true order, chalk, oolites, lias, red marls and coal, and the metalliferous rocks. But there are glimpses of such a law in earlier writers. Thus Woodward, in one of several false assertions (to which his inconsiderate anxiety in support of a favourite doctrine led him), speaks of chalk as an upper stratum, and accounts for this circumstance by reason of its lightness as compared with other rocks. He even accounts for the occurrence of Echinodermata, &c. in the chalk, while different exuviæ lie in other rocks, by the same rule of specific gravity, the lightest rocks having the lightest shells. This reasoning was ridiculous, but the passage is curious, as showing a certain amount of knowledge on a point which had been the least investigated.

Strachey's important memoirs embrace only a limited, though very interesting district, and are connected with a childish hypothesis. A paper by the Rev. John Holloway (Philosophical Transactions, 1723,) describes the course of the clay vales on the Cam, Ouse, Nen, and Isis, the ridge of sand-hills of Woburn, and the chalk of Gog-Magog and the Chiltern Hills, as parallel masses of strata, "confirming what Woodward had said of the regular disposition of the earth into like strata, or layers of matter common through vast tracts ;" and hence makes a question whether the fuller's earth, which is found near Woburn, only in the ridge of sand-hills, "may not probably be found in other parts of the same ridge of sand-hills among other like matter?" These memoirs following Lister's proposal for a ' Map of Soils,' (1684), and succeeded by Woodward's ' Natural History of Fossils,' seem to open a great portion of the subject of modern geology, and appear remarkably fitted to excite and assist inquiry in England. Yet, until the appearance of an ' Essay on the Cause and Phænomena of Earthquakes,' by the Rev. John Michell (Philosophical Transactions, 1760), there was nothing published of importance in the advancement of geology. The subject which Michell undertook to investigate, compelled him to look at the construction of the earth on a large scale, and to pass beyond the limits of Britain for illustrations. Accordingly we find him stating some classes of data, such as the general facts of stratification, the position of the strata, their horizontality under plains and acclinal rise toward mountain ridges, for the purpose of establishing a general conclusion, which is thus expressed:—

" From this formation of the earth it will follow, that we ought to meet with the same kinds of earths, stones and minerals, appearing at the surface, in long narrow slips, and lying parallel to the greatest rise of any long ridges of mountains; and so, in fact, we find them." He gives as illustrations the regions of the Andes and the Sierra in South America, and the mountain ranges of North America, and then adds, " In Great Britain we have another instance to the same purpose, where the direction of the ridge varies about a point from due north and south, lying nearly from north by east to south by west. At considerable distances from large ridges of mountains the strata for the most part assume a situation nearly level, and in the mountainous countries are generally formed out of the lower strata; so the more level countries are generally formed out of the upper strata of the earth."

The author of these just and admirable generalizations became, in 1762, Woodwardian Professor at Cambridge, and held that appointment till about 1770. He then accepted the rectory of Thornhill, near Dewsbury in Yorkshire, and in his journeys to and from Cambridge crossed the rocks between the " coal strata of Yorkshire " and the " chalk." By these journeys he became sufficiently acquainted with the series of secondary formations in England to assist with his knowledge both Smeaton and Cavendish. Smeaton wrote on the back of a letter bearing the London post-mark of November 21, 1788, the following words :—

" *Mr. Michell's Account of the South of England Strata.*

Yards of thickness.

Chalk	120
Golt	50
Sand of Bedfordshire	10 to 20
Northamptonshire lime, and Portland lime, lying in several strata	100

Yards of thickness.

Lyas strata . . . . 78 to 100
Sand of Newark . . . . . about 30
Red clay of Tuxford, and several . . 100
Sherwood Forest pebbles and gravel . 50 unequal.
Very fine white sand . . . . . . uncertain
Roche Abbey and Brotherton limes . . 100
Coal strata of Yorkshire . . . . . ”

This section is a correct exponent of the series of strata really observable between Yorkshire and the country around Cambridge, except in regard to the (alluvial) "sand of Newark" and the "very fine white sand," which *seems to be a mistake, perhaps of position,* as sand which might be so described lies above the coal and below the Roche lime. It is also tolerably complete, *except in the oolites,* which were never analysed till Smith began his career of discovery.

The information which Michell possessed must have produced a great influence on the progress of positive geology had he retained the Woodwardian Professorship. Perhaps however, in this case, he might never have had the opportunity to gather the knowledge which he so freely distributed. That the public interest in such matters was very slight, is evident from the facts that Smeaton's Memorandum remained among his unpublished papers till 1810, when accident revealed it to Mr. Farey; that Whitehurst (publishing in 1777) quotes Michell's paper on earthquakes, but never alludes to his or any other scale of stratification; and that no British or foreign writer, not even a Woodwardian professor in the eighteenth century, published, employed, or in any manner alluded to, any such list!

Whitehurst, though inferior to Michell in philosophical power, and not possessed of the same firm and appropriate idea of the earth's structure, has the merit of announcing more distinctly than any previous writer the law of the settled order of succession among the strata, which must ere this have been rather generally allowed. "The arrangement of the strata," he tells us, "in general is such, that they invariably follow each other, as it were, in alphabetical order, or as a series of numbers, whatever may be their different denominations.......The strata," he observes, "follow each other in a regular succession, both as to thickness and quality, insomuch that, by knowing the incumbent stratum, together with the arrangement thereof in any particular part of the earth, we come to a perfect knowledge of all the inferior beds, so far as they have been previously discovered in the adjacent country."

Whitehurst distinguishes between stratified and unstratified rocks, and notices the effects of faults, but his work is clouded in a formal cosmogony.

Smeaton, interested about water-cements, and finding the lias of Somerset to furnish lime of the required quality, took some pains to ascertain the places where such lias occurred. He did not actually trace its course, but makes this statement:—

"In travelling from Glamorganshire through Monmouthshire, Gloucestershire and Warwickshire into Leicestershire, I found such frequent instances of ordinary walls and cottages built with stone that appeared to me to be blue lias, the mortar also being of the same hue, that I have not a doubt but that the curious naturalist, in making this expressly an object of search, would be able to trace it from Aberthaw and Watchet quite to Barrow, though probably with several breaks, as is usual in the arrangement of strata in the earth. From Leicestershire it appears to pass by the Vale of Belvoir into Nottinghamshire and Lincolnshire, for a species of this kind of stone is used in some of the buildings about Newark ; and the Great North Road is repaired with the blue lias stone for a considerable length in the post stage between Newark and Grantham: at Long Bennington (a village of Lincolnshire through which the road passes on that stage) there is a limekiln for burning it. I have not yet seen it further north than this, nor anywhere north of the Trent." (Book iii. chap. iv. page 115.)

In a similar spirit, he conjectures that the water-setting lime of Petersfield may be found in all the range of chalk hills from Lewes to Petersfield, and probably thence into Surrey to Guildford and Dorking. (Page 116.)

In what degree at this time (probably about 1786) Smeaton had been aided by communications from Michell does not appear. Michell is quoted by him, in page 117, as giving him information on another subject. In speaking of the Bath freestones, he does not connect them with those of Northamptonshire, nor does he employ any of the terms which a person possessed of Michell's views would have naturally chosen. It is probable that Michell's list was unknown to Smeaton till 1788, or perhaps later, and that he left the chapter on water-cements unaltered after 1787, which date is mentioned in it, though the publication did not take place till 1791. This is the more probable, as he quotes Mr. Cavendish and Dr. Blagden for information that the blue lias exists at Lyme Regis (p. 116). From this mere hint, who would have conjectured that about this time Blagden and Cavendish made extensive journeys in the south and west of England with the express purpose of determining the succession of the strata, and tracing their courses through the English counties? This fact, so important to the history of geology, I have learned by inspection of the unpublished Cavendish MSS., which some time since were placed by their noble proprietor in the hands of the Rev. W. V. Harcourt. The tour was evidently well planned; it was rather a tour of inspection and verification than of original investigation; the strata are spoken of familiarly, as things found where they might be expected, rather than as unknown objects of discovery. It is clear that the tour was consequent on, and planned with reference to, previous information; and the correspondence in the same collection proves satisfactorily the informant to have been John Michell! To this distinguished ornament of British science (*whose name is omitted in most of our biographies*!), we thus trace (1760) not only some of the grandest views in early geological science, but, *in later life* (before 1788), of the first approximate scale of the secondary strata of England That he never *published* this scale, may be accounted for by his advancing age and the freedom of his MS. communications to Cavendish and Smeaton, who might reasonably have been expected to work out his ideas, and secure for him the honour of originating them. But Smeaton died in 1792, Michell in 1793. Cavendish had discovered the composition of water, and was, besides his engagements in other branches of science, too fastidious a judge of his own performances to be induced to publish the results of a hasty journey. He died in 1810, two years after the formation of the Geological Society of London; but who ever heard from one who could have connected the fame of Michell and Sedgwick, and filled the most re-markable void in the history of English geology, a single word respecting his own or Blagden's inquiries?

We have now cleared the way to William Smith, who was born about the time when Michell quitted Cambridge, and acquired the first clear and consistent view of the series of stratification near Bath, just about the time when Smeaton published his chapter on water-cements. In a great degree self-educated, forced to struggle into notice in a laborious profession, unacquainted even with the names of Michell and Cavendish, the phænomena which had caught the notice of Strachey in Somersetshire fixed the attention of the young man. But instead of supposing with Strachey, strata curved from the centre to the circumference of the earth, or looking with Michell at the Andes and the Sierra, or with Whitehurst expanding into propositions the limited expe-rience of the miners of Derbyshire, he concentrates his at-tention on the general regularity of the strata near Bath (1790–91), broken only by the single case of unconformity between the red ground and the coal. He indulges no speculations of horizontal strata in plains and inclined strata in mountains, but seeing and proving as a local fact that the strata of Somerset have a general inclination to the east or south-east, turns all the energy of his mind to determine if a similar law applies in other districts, dwells for this purpose on every memory of his earliest years (1787–8–9), seizes every occasion of travelling which limited means per-mit, accepts with joy the opportunity of a long journey through England and Wales (1794), takes eager notes of every hill and every quarry, and returns satisfied that the surface of our island is formed on the edges of strata which are continuous for great areas, which succeed one another in a certain order, preserve approximately the same thickness and quality, produce similar soils, have similar uses, and affect in like manner the drainage, the elevation, the physical geography, and the whole aspect of the country. Once master of these ideas, he took them as the guiding-star, the one object of his life; illustrated them by models, maps and collections; deduced from them new methods of drainage, new principles of mineral surveying, new practices in engineering; and at length, after minute and repeated examinations, not only completed a Table of Stratification, and coloured maps, and arranged collections, such as never were conceived before, but arrived at further and more magnificent discoveries, of which scarcely the least indication can now, by the most scrutinizing search, be found in the records of earlier inquiry.

Accustomed to view the surfaces of the several strata which are met with near Bath uncovered in large breadths at once, Mr. Smith saw with the distinctness of certainty, that "each stratum had been in succession the bed of the sea;" finding in several of these strata abundance of the exuviæ of marine animals, he concluded that these animals had lived and died during the period of time which elapsed between the formation of the stratum below and the stratum above, at or near the places where now they are imbedded; and observing that in the successively-deposited strata the organic remains were of different forms and structures—Gryphites in the lias, Trigoniæ in the inferior oolite, hooked oysters in the fuller's earth,—and finding these facts repeated in other districts, he inferred that each of the separate periods occupied in the formation of the strata was accompanied by a peculiar series of the forms of organic life, that these forms characterized those periods, and that the different strata could be identified in distant localities and otherwise doubtful cases by peculiar imbedded organic remains.

Of this important series of inferences, which is now the accepted basis of the natural history of the earth,—the record of its chronology, hardly the least trace or foreshadowing appears in the writings of earlier explorers. A single passage in Lister, who recognised multitudes of a small belemnite in the Wolds of Yorkshire and Lincolnshire, "at semper in terrâ rubrâ ferrugineâ," only makes us regret that so curious an observation should have been clouded with doubt or disbelief of the animal origin of these fossils,—a state of mind which effectually prevented the progress toward the acquisition of a great truth which he seemed well fitted to make.

Woodward's strange assertion (1696), that the matter of the earth fell from suspension in a liquid, so that the strata are laid one upon another in the order of their specific gravity, led him necessarily to maintain the further error, that the shells which lay in these strata had the same specific gravity; the heavier shells (as "*Conchæ, Pectines* and *Cochleæ*") lying in the heavier and lower strata of sandstone, the lighter shells (as *Echini*) lying in the lighter and upper strata of chalk; and organic bodies still lighter (" as shells of lobsters and crabs, bones and teeth of fishes "!) would subside last of all, and lie at or near the surface. This opinion, though it be in fact a mass of errors, shows, what indeed cannot be imagined to have happened otherwise, that an observer who was singularly exact in noticing the localities of his fossils, was led to perceive, though very dimly, that their distribution in the earth was subject to *some law*, but he met with no success in his attempt to conjecture *what the law might be*.

Whitehurst notices in the Derbyshire hills the distinctness of the shelly contents of the limestones from the vegetable remains of the coal series, but the view with which this very obvious circumstance is associated, shows that he had no right apprehension of any general law of the distribution of fossils. Some foreign writers, in particular Scilla and Rouelle, appear to have made very just comparisons of the natural associations of fossil shells, corals, &c. in the earth, with the groups of similar objects as they are found in the

sea, and thus to have produced new proofs of the organic
origin of these fossil bodies; but they give no sign of any
knowledge of the *limitation of particular tribes of organic
remains to particular strata,* of the *successive existence of
different groups of organization,* on *successive beds of the
antient sea.* Mr. Smith's claim to this happy and fertile
induction is clear and unquestionable: the process by which
he arrived at it is perfectly known; the proof of its truth and
value grows brighter from day to day, as the philosophy of
geology advances; and it is gratifying to remember that the
wreath which he won was placed on his brow by men who
had gained honour in following his footsteps, and were neither slow nor reluctant to acknowledge the obligation.

Possessed of this clue, Mr. Smith walked securely through
the labyrinth of the strata, classing them in one district and
rediscovering them in another, till the Map of the Island
grew in his hands to be fit for publication, at a time (1800)
when, in the mass of English society, the strata were unknown and geology had no name. By the same guide he
was conducted to a clear separation of the superficial deposits of gravel, sand, clay and peat—often enclosing the
remains of quadrupeds and rolled specimens of shells which
had been drifted from other situations, and always indica-
ting surface-agencies in their accumulation—from the sandstones, limestones and shales of older date, due to marine
action, and enclosing marine exuviæ in the place and under
the circumstances of their original deposition.

Thus originated (previous to 1796) that distinction of
diluvial and stratified deposits which in after years afforded
to Buckland the theme of a splendid theoretical effort; a
distinction which still retains its importance, and still gives
occasion for new speculations as the circle of facts widens,
and the phænomena of diurnal nature are more carefully
compared with the primæval monuments of the globe.

In the study of the stratified, diluvial and alluvial deposits, with their characteristic organic remains, and in the
practical applications of these subjects to agriculture and
mining, Mr. Smith exhausted all the resources of his mind:

L

he was very little acquainted with some other important parts of geological research. The granitic and other igneous rocks he rarely sought opportunities to examine; the general history of the ancient effects of subterranean heat, and the modern phænomena of volcanoes and earthquakes, interested him rather as a spectator than as a student : in geological dynamics he seized many leading generalizations in the configuration of the terraqueous surface which seemed to depend on the operation of the atmosphere, rivers, and the sea, but left almost untouched the questions of ancient climate, the changes of ocean-level, the elevation of land, the mechanical laws of fracture, and the chemical laws of venigerous fissures. In the promotion of these and other high branches of geological research, other names, both of earlier and later date, must stand on tablets separate from that of William Smith. Even in his own well-laboured field, younger cultivators are rearing richer crops with better hopes of a propitious harvest :—were he alive who first reclaimed the waste, no voice would be louder in their praise, no heart throb with higher delight at every fresh triumph of their toil.

APPENDIX.

———————

PAGE 1.—In this opinion Mr. Smith was probably mistaken; but he had collected some information to elucidate the point, which the Editor has not found either leisure or opportunity to complete, by searching the parish records of Idbury in Gloucestershire, and the registers of wills at Oxford. It is besides of little consequence to the history of "Strata Smith" from what *gens* he remotely sprung; his immediate ancestors and all his connexions were in humble life; on the oolitic soils which they had cultivated for ages he was born and bred; on these he planted, in advance of all other men, the standard of geological discovery; to the study of these his last days of active mind were given; and in these, according to a natural, if fanciful wish, his last remains are laid to rest.

Page 15.—The old farm-house in which Mr. Smith lived, while examining the strata of the coal-fields about High Littleton, was called Rugborne. Into his own diet there, milk entered largely, but the honest and hospitable farmer preferred the rich cider of that fertile vale; and Mr. Smith, when a sexagenarian, used to laugh with boyish glee at the statement of one of the farmer's men, that " he allowed his master four hogsheads of cider a-year for his own drinking." The quantity of "drink" really consumed by farm labourers in some parts of Somersetshire and Herefordshire during harvest is almost incredible.

Page 17.—The Swan Inn still exists at Dunkerton, and is a convenient station for any one who wishes to explore the geological structure of one of the two valleys in which Mr. Smith measured the eastward declination of the oolitic strata. Its doors are *now* thronged by innumerable coal carts; the interior was *then* the resort of the engineers, contractors and " navigators" engaged on the Coal Canal.

Page 30.—Soon after the publication of the great "Map of the Strata of England and Wales," Mr. Smith found the means to represent with additional exactness the subdivisions of the oolites, and drew up a geological table somewhat more extended than that which served as an Index to his Map. Thus has he given *three tables* which deserve more than a passing glance of comparison, because the causes of their difference and the sources of the successive improvements are perfectly known to the Editor, partly by Mr. Smith's communications, and partly by tracing with his maps in the hand the lines of country where his observations were made.

The following is a Comparative View of the Names and Succession of the Strata given in these Tables.

Table drawn up in 1799.	Table accompanying the Map, drawn up in 1812.	Improved Table drawn up in 1815 and 1816, after the first copies of the Map had been issued.
	London Clay..............	1. London Clay.
	Clay or Brick-earth	2. Sand.
		3. Crag.
	Sand or light Loam	4. Sand.
1. Chalk	Chalk.....................	5. Chalk { Upper. Lower.
2. Sand	Green Sand	6. Green Sand.
	Blue Marl..................	7. Brick-earth.
		8. Sand.
	Purbeck Stone, Kentish Rag and Limestone of the vales of Pickering and Aylesbury	9. Portland Rock. 10. Sand. 11. Oaktree Clay. 12. Coral Rag and Pisolite.
	Iron Sand and Carstone ..	13. Sand.
3. Clay..................	Dark blue Shale	14. Clunch Clay and Shale.
		15. Kelloway's Stone.
	Cornbrash	16. Cornbrash.
4. Sand and Stone........		17. Sand and Sandstone.
5. Clay................		
6. Forest Marble	Forest Marble Rock	18. Forest Marble.
		19. Clay over Upper Oolite.
7. Freestone	Great Oolite Rock	20. Upper Oolite.
8. Blue Clay .... 9. Yellow Clay .... 10. Fuller's Earth ...... 11. Bastard ditto and Sundries		21. Fuller's Earth and Rock.
12. Freestone	Under Oolite	22. Under Oolite.
13. Sand		23. Sand.
		24. Marlstone.
14. Marl Blue	Blue Marl............	25. Blue Marl.
15. Blue Lias	Blue Lias	26. Blue Lias.
16. White Lias	White Lias	27. White Lias.
17. Marlstone, Indigo and Black Marls.........		
18. Red-ground	Red Marl and Gypsum	28. Red Marl.
19. Millstone	Magnesian Limestone	29. Redland Limestone.
	Soft Sandstone	
20. Pennant Street....... 21. Grays .... 22. Cliff................. 23. Coal................	Coal Districts	30. Coal Measures.
	Derbyshire Limestone	31. Mountain Limestone.
	Red and Dunstone..........	32. Red Rhab and Dunstone.
	Killas, or Slate	33. Killas.
	Granite, Sienite and Gneiss {	34. Granite, Sienite and Gneiss.

In these tables we see typified unequal degrees of knowledge, different estimates of the value of groups, but the same independent mind wielding its own resources uninfluenced by the progress of opinion in other men. The first table represents, with almost entire exactness, the natural section on the road from Warminster to Bath. It contains no tertiary strata: of the existence of such farther to the east Mr. Smith was perfectly aware, but in accordance with his scheme of notation mentioned on page 22, the chalk, as forming "a grand feature" and perfectly known, is marked No. 1. On the road in question the greensand No. 2 is perfectly seen; but Mr. Smith's section notices no strata between this and the clunch or Oxford clay. I had always supposed this omission of all mention of Kimmeridge clay and coralline oolite to be a mark of the incompleteness of his knowledge at the time, but the supposition only proved my own ignorance. Lately, on a careful inspection of the ground, Sir H. De la Beche and myself have found that Mr. Smith was perfectly right in leaving without notice these beds, for they are entirely absent on that line of section, being covered up by the over extension of the greensand so as to touch the Oxford clay.

Soon after the table was drawn up, Mr. Smith discovered the coralline oolite (which he then called coral rag or superior oolite) in the park at Longleat, and noticed the superjacent blue or oaktree clay (now called Kimmeridge clay), a fact which Sir H. De la Beche and myself have lately recognised with pleasure and advantage.

Below the clay No. 3 should have been found some distinct mention of the cornbrash, which it is almost inconceivable that Mr. Smith could have overlooked; it is probably included as a member of the series No. 4. From this point to the coal formation the section is remarkably good and complete. The term marlstone was afterwards transferred to some ochraceous beds beneath the sand No. 13. No strata are named below the coal, evidently because they were not *accurately* known.

In Feb. 1802, he began to acquire correct general views of the upper oolitic groups, with their associated clays and sands (see p. 41), which were not well exhibited in the "unconformable" country south-east of Bath; yet he was not perfectly satisfied about them in 1812, when the principle of the colouring of his map was arranged; and hence arose two evils, an indecisive delineation of the areas of these rocks, and an incomplete enumeration of their constituent groups in the memoir accompanying the map. In that memoir the tertiary strata are in three groups; the Golt finds a place, but the Portland and Oxford oolites, and their accompaniments, were confused in groups very unsatisfactory to Mr. Smith. This confusion he perfectly cleared away almost immediately after the preparation of the first copies of his Map, by a careful examination of the Vale of North Wilts, and a comparison of the sections of Swindon with those of the Vale of Aylesbury, Vale of Wardour, and Isles of Portland and Purbeck. The oolitic groups of Bath are mentioned in the memoir only in general terms, corresponding to the colours employed on the map. The strata below the coal are added upon the same restricted principle.

The third table exhibits the series of strata in England and Wales, nearly as we accept them at the present day. In this table the analysis of the successive marine groups, from the chalk to the lias inclusive, is so remarkably judicious, and the formation of the groups so nearly perfect, that it is become the standard by which all our best maps and sections are measured; and, if inscribed on the tomb of " Strata Smith," would as appropriately mark a great conquest in natural science as the long array of figures on the monumental stone declares the triumph of the mathematician.

Page 77.—The following is a general list of the publications on the Geology of England and Wales, by William Smith :—

" A Geological Map of England and Wales, with part of Scotland ; exhibiting the Collieries, Mines, and Canals, the Marshes and Fen Lands originally overflowed by the Sea, and the varieties of Soil, according to the variations of the Substrata; illustrated by the most descriptive Names of Places, and of Local Districts; showing also the Rivers, Sites of Parks, and principal Seats of the Nobility and Gentry, and the opposite Coast of France." By William Smith, Mineral Surveyor. The Map is engraved on a scale of five miles to an inch, and consists of fifteen large sheets. Commenced in 1812, published 1815. Size, 8 feet 9 inches high, by 6 feet 2 inches wide.

In sheets, with Memoir, 5l. 5s.
On canvas and rollers, 7l.; varnished, 8l.
In case for travelling, 7l.; on spring rollers, 10l.

" A Reduction of Smith's large Geological Map of England and Wales, exhibiting a general View of the Stratification of the Country; intended as an Elementary Map for those commencing the Study of Geology." 1819.

Price, in sheet, neatly coloured and shaded, 14s.

Mounted in case for Travelling, or on rollers, 18s.

" A New Geological Atlas of England and Wales, on which are delineated, by Colours, the Courses and Width of the Strata which occasion the varieties of soil; calculated to elucidate the Agriculture of each County, and to show the Situation of the best Materials for Building, Making of Roads, the Constructing of Canals, and pointing out those Places where Coal and other valuable Materials are likely to be found." By William Smith, Author of the Geological Map of England and Wales.

Part I. contains Norfolk, Kent, Wilts and Sussex. Price 1*l.* 1*s.* 1819.

Part II. contains Gloucester, Berks, Surrey and Suffolk. Price 1*l.* 1*s.* 1819.

Part III. contains Oxford, Buckingham, Bedford and Essex. Price 1*l.* 1*s.* 1820.

Part IV. containing a Map of the County of York, on four sheets. Price 1*l.* 1*s.* 1821.

Part V. containing Leicester, Nottingham, Huntingdon and Rutland. Price 1*l.* 1*s.* 1822.

Part VI. containing Northumberland, Cumberland, Durham and Westmoreland. Price 1*l.* 1*s.* 1824.

⁎ These Maps may be had separate, price 5s. 6d. each. Other Parts to complete this Work were left by Mr. Smith in a state of forwardness.

" A Geological Table of British Organized Fossils, which identify the Courses and Continuity of the Strata." By William Smith. Coloured, price 1s. 6d. 1815.

" A Geological Section from London to Snowdon; showing the Varieties of the Strata, and the correct Altitude of the Hills." By William Smith. Coloured, one sheet, price 7s. 1819.

" Geological View and Section of Norfolk, and through Suffolk to Ely." By William Smith. Coloured, one sheet, price 5s. 1819.

" Geological View and Section of the Strata through Hampshire and Wiltshire to Bath." By William Smith. Coloured, price 5s. 1819.

" Geological View and Section in Essex and Hertfordshire, and of the Country between London and Cambridgeshire." By William Smith. Coloured, price 5s. 1819.

"Geological View and Section of the Country from London to Brighton, through Lewes." By William Smith. Coloured, price 5s. 1819.

"Geological View and Section through Dorsetshire and Somersetshire to Taunton, on the road through Yeovil to Wimborn Minster," &c. By William Smith. Coloured, price 5s. 1819.

"Strata identified by Organized Fossils." Coloured Plates, 4to. Seven parts proposed, four published, 7s. 6d. each; commenced 1816.

"A Stratigraphical System of Organized Fossils;" compiled from the original Geological Collection deposited in the British Museum, with coloured Tables of the Geological Distribution of the Groups of Echinodermata. By William Smith. 4to. 1817. 15s.

THE END.

Printed by Richard and John E. Taylor, Red Lion-court, Fleet-street.

Printed in the United States
By Bookmasters